Georgiy S. Beloglazov

Organic inhibitors of corrosion of metals: quantum chemical study

Georgiy S. Beloglazov

Organic inhibitors of corrosion of metals: quantum chemical study

Also: inhibitors of hydrogen absorption by steel

LAP LAMBERT Academic Publishing

Impressum / Imprint

Bibliografische Information der Deutschen Nationalbibliothek: Die Deutsche Nationalbibliothek verzeichnet diese Publikation in der Deutschen Nationalbibliografie; detaillierte bibliografische Daten sind im Internet über http://dnb.d-nb.de abrufbar.

Alle in diesem Buch genannten Marken und Produktnamen unterliegen warenzeichen-, marken- oder patentrechtlichem Schutz bzw. sind Warenzeichen oder eingetragene Warenzeichen der jeweiligen Inhaber. Die Wiedergabe von Marken, Produktnamen, Gebrauchsnamen, Handelsnamen, Warenbezeichnungen u.s.w. in diesem Werk berechtigt auch ohne besondere Kennzeichnung nicht zu der Annahme, dass solche Namen im Sinne der Warenzeichen- und Markenschutzgesetzgebung als frei zu betrachten wären und daher von jedermann benutzt werden dürften.

Bibliographic information published by the Deutsche Nationalbibliothek: The Deutsche Nationalbibliothek lists this publication in the Deutsche Nationalbibliografie; detailed bibliographic data are available in the Internet at http://dnb.d-nb.de.

Any brand names and product names mentioned in this book are subject to trademark, brand or patent protection and are trademarks or registered trademarks of their respective holders. The use of brand names, product names, common names, trade names, product descriptions etc. even without a particular marking in this works is in no way to be construed to mean that such names may be regarded as unrestricted in respect of trademark and brand protection legislation and could thus be used by anyone.

Coverbild / Cover image: www.ingimage.com

Verlag / Publisher:
LAP LAMBERT Academic Publishing
ist ein Imprint der / is a trademark of
AV Akademikerverlag GmbH & Co. KG
Heinrich-Böcking-Str. 6-8, 66121 Saarbrücken, Deutschland / Germany
Email: info@lap-publishing.com

Herstellung: siehe letzte Seite /
Printed at: see last page
ISBN: 978-3-659-39656-4

Zugl. / Approved by: Dodoma (UDOM), Tanzania

In loving memory of Dr. Mikhail I. VAKHRIN, associate professor at Perm State Pharmaceutical Academy (PGFA) who was my teacher, colleague, and leader (Head of Department of Physics and Mathematics)

Georgiy S. Beloglazov

ORGANIC INHIBITORS of CORROSION of METALS and HYDROGEN ABSORPTION by STEEL: QUANTUM CHEMICAL APPROACH to EXPLAINING THEIR ADSORPTION and PROTECTIVE ACTION

Department of Physics, the University of Dodoma (UDOM). P.O. Box 259, Tanzania

Perm (PGFA) – Dodoma (UDOM – Tanzania), 2013

ABBREVIATIONS

CR –   corrosion rate

ECISRB –   efficiency of corrosion inhibition in media inoculated by SRB

ECI –   efficiency of corrosion inhibition (in sterile media)

EI –   inhibition efficiency

EIHA –   efficiency of inhibiting hydrogen absorption (by steel)

ESRB –   efficiency against sulphate reducing bacteria

ExD –   experimental data

HOMO –   highest occupied molecular orbital

IC –   inhibitor of corrosion (of metal/s)

IH –   inhibitor of hydrogen absorption (by steel)

LUMO –   lowest unoccupied molecular orbital

MNDO –   modified neglection of differential overlapping [60, 61]

OIn –   organic inhibitor (i.e., IC and IH together)

PM/3 –   parametrization method 3 (quantum chemistry [25, 60, 61])

QCC –   quantum chemical characteristics

QCD –   quantum chemical descriptors

SRB –   sulphate reducing bacteria

WHDC –   water-hydrocarbon distribution coefficient

3

ACKNOWLEDGEMENTS

I would take this opportunity to thank kindly my wife Lyudmila and children for making coffee and a lot of other good things for me. Also, I need to thank Ms. Alison Whitehead for her indispensable kind help in proofreading the book.

Many kind thanks to my English teachers: Mrs. Olga S, Skvortsova, Mrs. Svetlana D. Gnusina, Mrs. Tamara A. Alikina, Mrs. Olga V. Ukholva, Dr. Galina P. Bazhina, Dr. Elena D. Igoshina, Dr. Olga V. Kamenskikh, Dr. Warshavskaya.

INTRODUCTION

The annual losses due to corrosion are reported [71] to be as much as \$276 billion (2010) only in one of the countries of the world. Nowadays, use of inhibitors is said to be the most promising way of protecting metals from corrosion [1–3] as well as steel from hydrogen absorption [4]. More and more efficient, cost-efficient (inexpensive) and environmentally- friendly inhibitors appear on the daily basis [1–4, 8, 29, 33, 34, 53]. Adsorption of organic compounds acting as corrosion inhibitors (IC) of metals or/and as inhibitors of hydrogen absorption of steel (IH) is regarded as the main mechanism of the first stage of protection action of, including other, organic inhibitors (OIn) [1–4]. The important role during the adsorption process is played by polar groups (such groups as OH, SO_2, SO_2NH_2, SO_2NHR, SO_2NR_2, $-NH_2$) and heteroatoms (such as N, S and O) [5–10]. Currently, adsorption of organic molecules on metals is studied not only experimentally but also theoretically, for example [11, 12, 54–57]. With the help of high efficient IH, it is possible to prevent the loss of mechanical properties of high strength steels widely used in modern techniques. Highly effective ICs and IHs are in demand to prevent any accidental destruction as a result of the practice of etching of steel after its thermo-mechanical treatment.

Thanks to significant developments in computer technologies, now it is possible to replace some of the expensive and time-consuming technical experiments (including organic synthesis of ICs and IHs) for various application media and conditions with computational simulations, provided the capacity of modern computers enables us to solve complicated problems such as the tasks of quantum chemistry where the operations on high order matrices are required in order to perform calculations of the adsorptive properties of actual multi-atomic systems. Just as an example, the opportunity of modelling adsorption of molecules containing up to 80 atoms (20…40 of which are representing metal adsorbent within a cluster «adsorbed OIn + adsorbent») even using the most well defined (and thus the most complicated) quantum chemical methods [60, 61] (i.e. *ab initio* methods) is now feasible [13–24]. The purpose of such a «digital experiment» is purposeful organic synthesis of efficient inhibitors of corrosion and hydrogen absorption of metals on the basis of known qualitative and quantitative structure-to-property relationships. On the first stage of such modelling, it is worth finding out: (1) which quantum chemical descriptors (QCD) being obtained from the computations prove to be in a significant (and at best if one-to-one) relationship with experimentally gained data on OIn efficiency against corrosion and hydrogen absorption, (2) the right size of polyatomic cluster which, on one hand, already correctly maps the qualitative and quantitative dependencies characterising the important processes involving metal protection (including adsorption of OIn on the metal surface being protected) and their results (taking into account media composition and other conditions of corrosion and/or hydrogen absorption), on the other hand, a researcher is not yet exposed to such massive problems as too long calculations.

1.1. Quantum chemical study of adsorption of organic inhibitors of corrosion on Al surface

Below are given some examples of quantum chemical modelling applied for different metal adsorbents (such as Al and Cd) at various conditions of surface (oxide, hydroxide, pure metal – the latter practically justified for most cases of natural or industrial electrolytes and other solutions particularly at electrochemical processes).

Table 1

Correlation coefficients (%) between experimentally measured efficiencies of corrosion inhibition (ECI) and theoretically computed QCDs when modelling adsorption of OIn on Al

OIn concentrations (mMol/L)	Correlation coefficients (%) for the following three quantum chemical descriptors (QCD)		
	Dipole moment p_0 of OIn isolated molecule	Most probable angle $\alpha°$ between the changed dipole moment of OIn molecule and normal to Al_{20} surface	Relative change of module of dipole moment Δp_{ad} of OIn molecule when adsorbed on Al_{20} modelling cluster
1	–55 %	–46 %	0 %
5	–67 %	–58 %	21 %
10	–68 %	–67 %	30 %
15	–74 %	–73 %	37 %

In Table 1, some quantum chemical characteristics QCD computed using the restricted Hartree-Fock *ab initio* (STO-3G) method implemented in the software [25] are compared to the experimental data obtained for the same OIn (six various substituted phenols).

When computing the most probable angle between the vector of dipole moment of adsorbed OIn molecule and the normal to flat metal surface (see Table 1) the fact that dipole moment of OIn molecule is changed (modified) at adsorption of that molecule, is taken into account.

1.2. Quantum chemical study of adsorption of organic inhibitors of hydrogen absorption by steel on the surface of its electrodeposited Cd coating

Some results for five different substituted diantipirylmethanes («D1…D5» as OIn in the role of inhibitors of hydrogen absorption by steel by cadmium electroplating) obtained using the same quantum chemical method i.e. restricted Hartree-Fock *ab initio* (STO-3G) (program [25]) are listed in Table 2.

Table 2

Values of EIAC and QCD for OIn «D1…D5» (model cluster: Cd_{22})

OIn Title	EIAC (%) of the studied inhibitors of steel corrosion by cadmium plating at concentrations of OIn, mMol/L (numerator – D_k=1 A/dm^2, denominator – D_k=3 A/dm^2):			Computed dipole moment of OIn molecule, D	Electric charge summed over all atoms of adsorbed OA molecule when adsorbed on Cd_{22} modelling fragment ($\Delta Q_{ад}$, |e|)	Absolute change in the module of dipole moment of OA molecule when adsorbed on Cd_{22} modelling fragment (Δp_{ad}, D)
	1	2	5			
D1	78 / 44	81 / 54	84 / 56	1.65	−2.356	5.532
D2	78 / 46	84 / 58	81 / 58	14.90	−2.192	7.456
D3	84 / 51	86 / 63	86 / 63	10.56	−2.402	1.816
D4	84 / 56	89 / 66	89 / 66	8.85	−1.955	2.580
D5	84 / 61	89 / 71	89 / 68	19.19	−2.173	−0.889

Table 3

Pair correlation coefficients between EIAC and QCD

OIn concentration, mMol/L	Current density D_k, A/dm^2	Quantum Chemical Characteristics (QCD)		
		Dipole moments of isolated OIn molecules (p_o)	Electric charge summed over all atoms of adsorbed OA molecule when adsorbed on Cd_{22} modelling fragment (ΔQ_{ad})	Absolute change in the module of dipole moment of OA molecule when adsorbed on Cd_{22} modelling fragment (Δp_{ad})
1	3	63%	52%	−89%
1	2	81%	64%	−70%
2	2	72%	65%	−76%
5	2	69%	48%	−89%
5	3	60%	51%	−89%

Concentration dependences of the found correlation coefficients between expe-rimentally measured efficiencies of inhibitive action and theoretically computed values of the most probable angle between the vector of dipole moment (please note also the remark to Table 1) of OIn molecule as a result of adsorption on the surface of Al alloy and the surface normal of metal being modelled, are simbate in all three studied cases of condition of such surface, namely Al_{20}, $\{4Al_2O_3\}$ and $\{4Al(OH)_3\}$. Comparison of the pair correlation coefficients given in Tables 2 and 3 gives the evidence that mechanisms of protection of metal against corrosion and hydrogen absorption are different in case of Cd plating of steel and Al (as chief component of some industrial duralumin alloys).

Previously, very important results have been obtained by Larkin and Rosenfeld [54–57] by means of their own approach to a use of quantum chemical computations of the polyatomic molecular systems comprising (a) one or two molecules of an inorganic or organic corrosion inhibitor, and (b) cluster imitating a piece of surface of metal being protected (in various oxidation states).

The obtained results confirming the special role of dipole moment of OIn molecules agree with previously published data [17–24]. Generally the development of the proposed approach promising in the framework of the prospects of building possibly universal model capable of describing inhibitive action against corrosion and hydrogen absorption by metals in different media and under various conditions.

2. Understanding mechanisms of inhibiting action of N-containing organic compounds on corrosion of metals and hydrogen absorption of steel using quantum chemical computations

2.1. Study of substituted amines and amides

Empiric (Simple Huckel) [60] quantum chemical computations were used to account for the influence of adsorption of organic species chemisorbed on metals (as Fe, Co, Ni, Ru, Pd, Rh) on oscillation frequencies of their C–O, C–N, and C–C molecular fragments in the 1960ies [39] probably for the first time.

Also during the 1960s, Hansch [30] developed the analysis of relationships between hydrophobic, electronic or sterical properties of the molecules and biological functions. In the 1970s, Stuper et al. [44] developed an approach to such a relationship based on manipulations with formal structural (topology) descriptors as atoms distribution in molecules of interest by elements, different bonds by types (like C–N, C–O or C–S etc.) and multiplicity (like single, aromatic, double, triple and so on). According to [44], such formal descriptors are, on one hand, sufficiently characterising the properties under question and on another, are universal enough to be obtained for any chemical structure which is naturally so. Vigdorovitch & Tsygankova [12] thereafter have extensively developed this approach for essentially more complicated objects.

In publications [20, 21, 38, 66] and others, the author has studied the electron charge distribution using PMR (NMR on ^{1}H) in order to quantitatively identify the active adsorption sites of the free (isolated) molecules of inhibitors of corrosion of metals belonging to different organic classes.

To identify the active sites in the molecules of inhibitors, the author has researched the relationships between efficiencies of corrosion inhibition (ECI) and atomic charge distribution in molecules of the inhibitors. For that purpose, the author has selected several series of organic species that have been well studied as the inhibitors of corrosion and hydrogen absorption by steel, specifically in water-salt hydrosulfuric media. In table 4, the ECI values for a series of derivatives of diethylaminephenylacetic acid (as hydrochlorides) substituted both in aromatic ring and at N atom of amine group.

Table 4

Experimental data on ECI of the studied inhibitors

$$X-C_6H_4-Y-CH_2-N-(C_2H_5)_2 \cdot HCl$$

Substituents		ECI (%) at inhibitor concentrations (mMol · dm^{-3}):		
X =	**Y =**	**1**	**2**	**5**
H	−NH−CO−	50	60	47
o-CH$_3$		42	40	44
p-OCH$_3$		47	49	52
p-NO$_2$		38	41	47
o,o,p-(CH$_3$)$_3$		89	88	88
H	−CO−CH$_2$−	26	30	78
p-CH$_3$		60	61	62

All ECIs presented in the appropriate tables of the present work have been calculated on the basis of the data from experimental measurements (as proposed by professor Balezin [53]):

$$ECI\ (\%) = 100\ \%\ (K_o - K)\ /\ K_o, \qquad (1)$$

where K and K_o denote corrosion rates (CR) with and without an inhibitor, respectively.

K and K_o (on the basis of which the respective ECIs have been determined) have been measured experimentally [21] with steel specimen after their exposition to corrosion in Postgate B media [48, 49].

Methoxy group at aromatic ring acts on an active molecular site as donor of electrons stronger (especially at *para*-position to side chain) than methyl therefore inhibitors containing methoxy-groups exhibit more efficiency than the ones containing just methyl groups. Among homologous inhibitors, more efficient are those with more methyl groups.

The values of chemical shifts from high resolution NMR (PMR) spectra [21] are brought in Table 5. These are compared in Table 6 to the experimental data on ECI from Table 4 for corresponding species in order to establish correlation dependencies between those two rows of experimental results.

Chemical shifts δ are related to screening constants σ [50, 51] by means of the following equations:

$$\delta = 10^6 \cdot (H_o - H_{loc})/H_o \ (ppm), \qquad H_{loc} = (1 - \sigma) \ H_o, \qquad (2)$$

where H_o– is constant magnetic field corresponding to the resonance condition (NMR) for internal standard, H_{loc} – is the local magnetic field at current chemically inequivalent location of the resonant nucleus (in case of PMR, proton).

Table 5

PMR chemical shifts of the inhibitors
$X–C_6H_4–Y–CH_2–N–(C_2H_5)_2 \cdot HCl$

Substituents:		PMR chemical shifts δ (ppm) at sites:			
X =	Y =	$–C_6H_4–$ (average)	$–CH_2–$	$–CH_2–$ $(–C_2H_5)$	$–CH_3$ $(–C_2H_5)$
H	–NH–CO–	7.61	4.32	3.47	1.49
o-CH$_3$		7.16	4.43	3.52	1.51
p-OCH$_3$		7.32	4.24	3.42	1.44
p-NO$_2$		8.10	4.38	3.52	1.51
o,o,p-(CH$_3$)$_3$		7.02	4.38	3.29	1.33
H	–CO–CH$_2$–	7.65	3.35	3.15	1.43
p-CH$_3$		6.29	3.86	2.14	1.06

Table 6

Linear approximation regressions and pair correlation coefficients **r** between ECI (Z) and PMR chemical shifts (δ) for the studied class of corrosion inhibitors

Proton-containing groups		**r** (%)	Linear regressions Z (δ, ppm)
$–C_6H_4–$		**78**	$Z_1 = (6.6 \ \delta + 25) \%$
$–CH_2–$		42	$Z_3 = (-2.4 \ \delta + 60) \%$
$–C_2H_5$:	$–CH_2–$	– 64	$Z_2 = (-5.3 \ \delta + 73) \%$
	$–CH_3$	– 47	$Z_4 = (0.7 \ \delta + 48) \%$

The σ dependence on the characteristics of charge distribution in molecules is rather complicated [50, 51], nevertheless the data presented in Table 6 are in favour of existence of strong correlation dependence between ECI and δ in the two cases out of four: for aromatic ring r_1=+78 % and for –CH$_2$– group (the one in ethyl group), r_2= –64%. The absolute values of pair correlation coefficients «ECI – δ» monotoneously decrease in the following sequence: $|r1| > |r2| > |r3| > |r4|$ (see Table 6) where the chemicallly inequivalent sites «r1» – «r4» are designated in the following way:

$$X–C_6H_4–Y–CH_2–N–(CH_2 – CH_3)_2$$

r1   r4   **r2**   r3

Hence in the studied series of organic IC there are two active sites: especially active is the aromatic site X–C$_6$H$_4$– (giving r_1=+78 %), and also N atom of central amine group (providing with r_2 = –64%), while for the studied class of the molecules correlation between ECI and δ at the end methyl protons of C$_2$H$_5$–groups (i.e. from the side of alyphatic radical of the molecules) seems to be absent.

Because according to (4) chemical shift is eventually defined by charge distribution in a molecule, it would be interesting to compare ECI straight to quantum chemical results. In Table 7, the correlation coefficients are brought between ECI and some quantum chemical charactesistics (QCC) of the molecules of the studied IH. MNDO [60, 61] was chosen as the method within the commercially available software [25].

Table 7

Pair correlations between ECI of
X–C$_6$H$_4$–Y–CH$_2$–N–(C$_2$H$_5$)$_2$ and QCC of their respective molecules

QCC	Correlation coefficients **r** (%) between ECI and QCC for the concentrations of the inhibitors (mMol/L):		
	1	2	5
HOMO energy	– 16	– 20	– 60
LUMO energy	– 12	– 17	– 22
$\Delta E = E_{LUMO} - E_{HOMO}$	**62**	**55**	**76**
$\eta = (E_{HOMO} + E_{LUMO})/2$	**– 61**	**– 55**	**– 77**
Electric charge (Q) on C atom of the only methyl group (at –Y–)	**– 56**	**– 59**	**– 66**

Sum of electric charges ΣQ (9) on C atoms of the aromatic ring $-C_6H_4-$	66	61	50
Net charge Q on N atom	26	31	77

In Table 7, such QCCs as HOMO (which characterizes the first ionization potential of the molecule according to Kupmans theorem [26, 31]) and LUMO (according to the same authors and theorem, it characterises electron affinity of the whole molecule) are presented [26, 31].

Also, forbidden gap magnitude which equals double 'absolute rigidity' of a molecule as a whole [31], probably being **far** from any sense of a bond/fragment rigidity (i.e. force coefficients/constants) in molecular oscillations

$$\Delta E = E_{LUMO} - E_{HOMO} \tag{3}$$

together with absolute electronegativity of molecule designated in the literature [31] as «η»:

$$\eta = (E_{HOMO} + E_{LUMO})/2. \tag{4}$$

In Table 8, the results on EIHA from experiment are defined (based on eq. (5)):

$$EIHA = Z_H = (n_{inh} - n)/(n_0 - n) \ [4, 28], \tag{5}$$

where n_0, n and n_{inh} are the plasticities of metal expressed in the average number of turns of wire samples they can stand in experiments before they crack out: n_0 stands for intact reference samples made of the same material as those exposed to hydrogen absorption (n), and n_{inh} is for those samples that were also exposed to electrochemical hydrogen absorption but with inhibitor (IH) respectively.

The details of the experiment are the following: the above said wire samples were cathodically polarized in water media containing 0,1 n H_2SO_4 + 5 mg · dm^{-3} H_2SeO_3 (as a promoter of absorpion of the cathodically evolving hydrogen [4, 28, 52]) at the current density of D_K = 10 mA · cm^{-2}. These experimental data are compared to the computed QCC in Table 9 (the pair correlation coefficients **r** are presented only for such QCC for which at least some of the found **r** values turn to be significant).

Table 8

Experimental data on efficiency of inhibitors $X–C_6H_4–NH–CO–CH_2–N–(C_2H_5)_2 \cdot HCl$ against of hydrogen absorption (EIHA) by steel

Substituents:	EIHA (%) at concentrations (mMol/L):			
X =	1	2	5	10
H	2.6	5.8	10.6	31.0
p-Cl	4.3	6.6	6.8	5.9
p-NO$_2$	1.7	3.4	4.3	5.5
p-CH$_3$	2.4	2.4	4.3	20.0
p-OCH$_3$	2.4	3.3	3.8	11.4

In Table 9, the charge differences of the type

$$\Delta Q_{CN} = Q\,(C) - Q\,(N) \tag{6}$$

characterise «bond polarities» (C – N bond is brought to your attention as an example; here the author has in mind C as the carbon atom in the methylene group marked as «r4» in a scheme above, before the Table 7).

The «polarity» of a separate chemical bond is proposed by the author similarly to «local dipole moment» with the only two differences: (i) that «differential polarity of an individual bond» is for the only two electric charges forming the bond, and (ii) these two charges together in most cases do not form an electrically neutral sub-system. Further, the author is introducing the term «differential polarity of an individual bond between a single atom and an atomic group (in the equation below, e.g., such atomic group in mind is C_6):

$$\Delta Q\,(C, \Sigma C) = Q\,(C) - \sum_{i=1}^{6} Q(C_i), \tag{7}$$

where

$$\Sigma Q = \sum_{i=1}^{6} Q(C_i) \tag{8}$$

is the sum of Mulliken net charges on C atoms in aromatic ring (the summation in (7) and (8) is over these 6 carbon atoms), and Q (C) is Mulliken net charge on the same single C atom as in (6).

Table 9

Pair correlation coefficients between EIHA of the studied inhibitors (IH) $X–C_6H_4–NH–CO–CH_2–N–(C_2H_5)_2$ and QCC of their molecules

QCC	Correlation coefficients **r** (%) between EIHA and QCC for the concentrations of IH (mMol/L):			
	1	2	5	10
HOMO	− 18	− 15	− 15	− 28
LUMO energy	− 10	− 21	− 13	− 22
$\Delta E = E_{LUMO} − E_{HOMO}$	− 42	− 48	− 78	− 79
ΔQ (C, ΣC) (7)	21	43	49	**78**
$\Delta Q_{CN} = Q(C) − Q(N)$ (6)	60	54	42	**81**
Q(C), the net electric charge Q on C atom of the only methylene group	30	9	21	49
Q (N), the net electric charge заряд Q on N atom	− 28	− 29	− 33	− 62

Statistical testing the hypothesis on pair correlation coefficient r = 0 at confidence factor of 0,95 shoes that for a series of 10 studied chemical species confidence level r′ is 61% which means that so-called competing hypothesis on non-zero correlation is significant at $|r| > r′ = 61$ % (hence the significant pair correlation coeffricients are shown in Tables 7 and 9 as bold).

From Tables 7 and 9, it appears that in the series of the derivatives of diethyamine-phenylacetic acid under consideration, both HOMO and LUMO exhibit significant correlation with neither ECI nor EIHA; what brings the significant values of correlation coefficients, is their linear combinations in the forms of either ΔE (3) and η (4). Were the correlation between LUMO and ECI (or EIHA) significant, one should conclude that at adsorption of a molecule of IC (or IH) on a surface of metal being protected, an extreme cases of donor-acceptor interaction takes place, i.e. donating by metal of an electron (or pair) to IC/IH molecule (or, opposite, in case of significant correlation of HOMO and ECI/EIHA – accepting electrons from inhibitor molecule, none of the two being unlikely the case).

The results shown in Tables 7 and 9 witness the difference in mechanisms of action of the inhibitors in case of hydrogen absorption from the corrosion case. It is necessary to note that linear combinations of the energies of the boundary orbitals (HOMO and LUMO) stated above in eq's (3) and (4), have been shown also by other researchers (for example, [59]) as the best so far QCDs suitable for describing other kind of molecular activity (namely, biological).

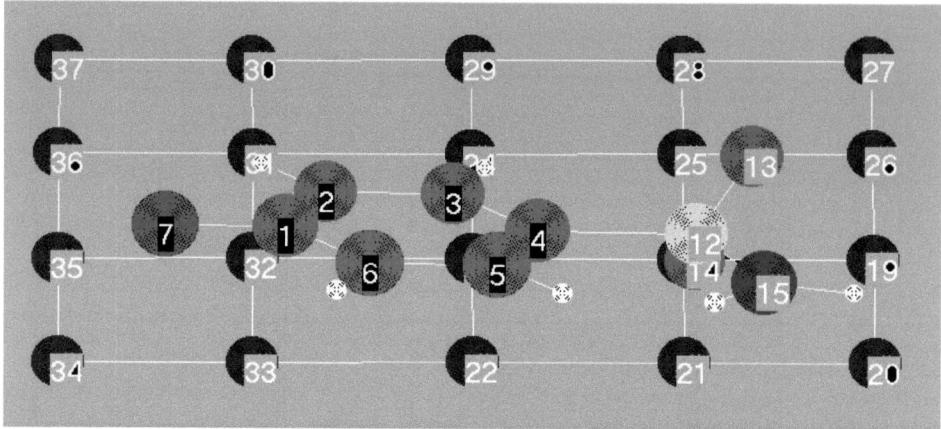

Fig. 1. "Plan" view on optimal geometry of p-F-C_6H_4-SO_2NH_2 adsorbed on Al_{20} cluster. Oxygen atoms have the numbers 13 & 14 (on the author's website [65] shown in red), Nitrogen atom is No. 15, Carbon atoms – 1…6, Fluorine is No. 7, Aluminium atoms are 19…37, the rest are hydrogen atoms.

In figures 1–7, are presented some results of quantum chemical computations obtained by means of using the programs [25] together with author's imaging software [58]. In figures 1 and 2, the proposed atomic numbering schemes (together with optimized geometries of some of the IC/IH molecules studied) are presented for adsorbed molecule of para-fluorobenzenesulphamide on the surface of a model cluster consisting of 20 Al atoms (in two projections). In Fig. 3, the proposed colour code seen on [65] website, is used to illustrate **re-distribution** of electron density in the same molecule occuring due to its adsorption on such model surface of the adsorbent. Various oxidation states of metal have been analysed; e.g., the adsorbate model shown in Fig. 3 corresponds to aluminium oxide Al_2O_3. In fact, Fig. 3 is a **<u>differential</u>** map giving some idea of absolute **changes** of the atomic net charges, in more detail available in ref. [65].

17

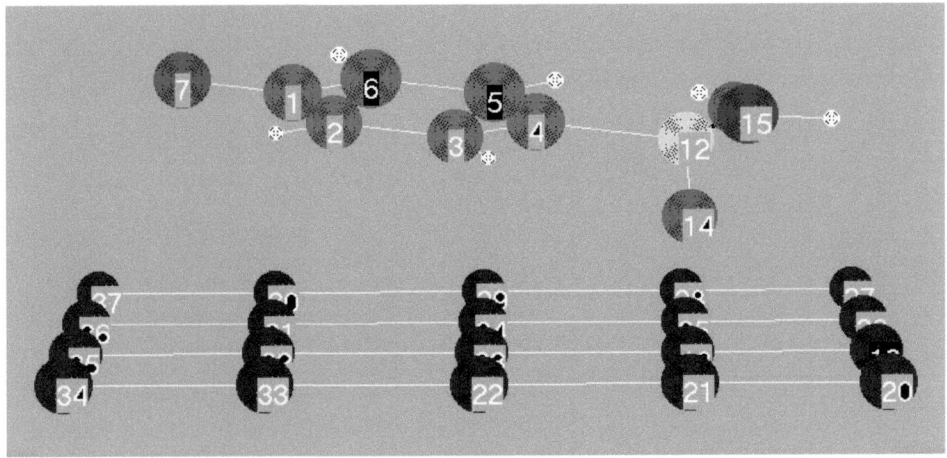

Fig. 2. "Isometric" view on optimal geometry of p-F-C_6H_4-SO_2NH_2
adsorbed on Al_{20} cluster. The same numbering as in Fig. 1.

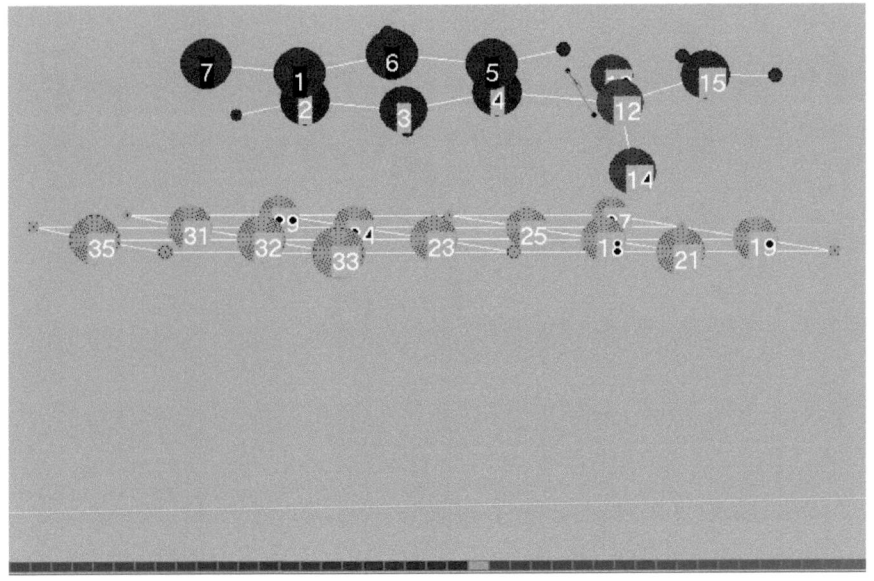

Fig. 3. Absolute **changes** of net atomic charges (*para*-fluorobenzenesulphamide) due to its
adsorption on model cluster emulating Al_2O_3 surface. The colour scale as such is in ref. [65].

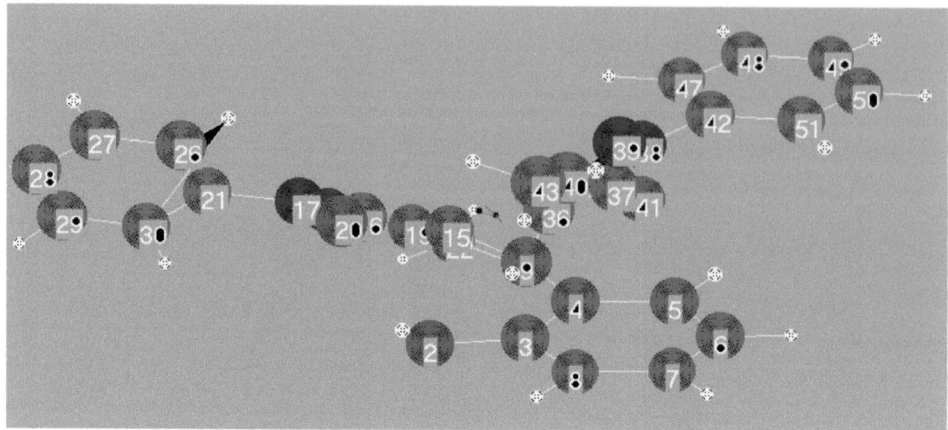

Fig. 4. Numbering scheme for oxyphenyl-diantipirylmethane molecule (for purpose of quantum chemical computations). The same colour code as in Fig. 1.

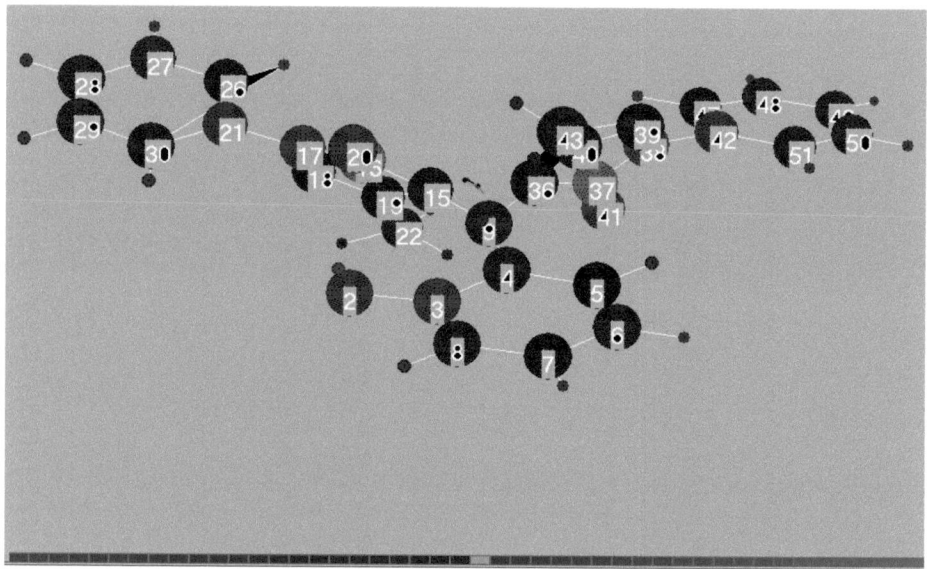

Fig. 5. Distribution of electric charge in an isolated molecule of oxyphenyl-diantipirylmethane (according to MNDO quantum chemical computation). The colour scale to be seen in [65].

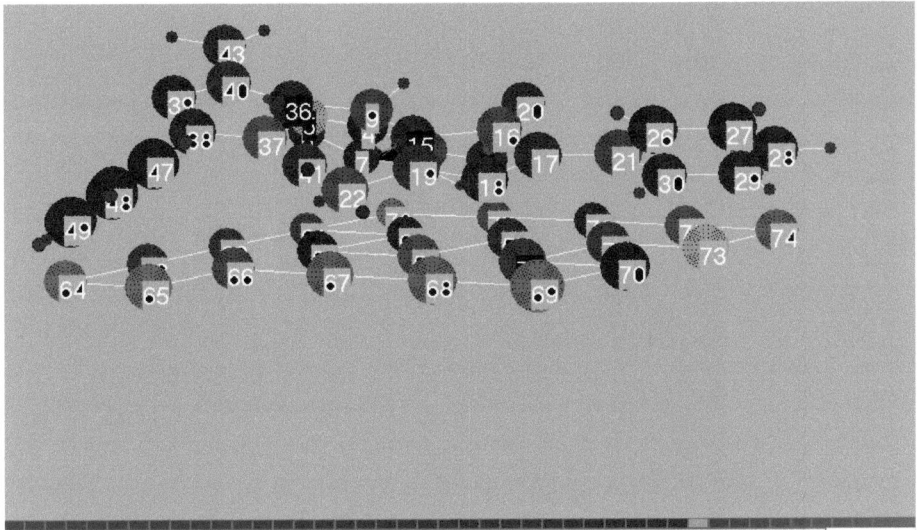

Fig. 6. Electric charge distribution in an isolated molecule of oxyphenyl-diantipirylmethane adsorbed on Al_{20} model surface (as a plane cluster) according to MNDO quantum chemical data. The colour scale is available in ref. [65].

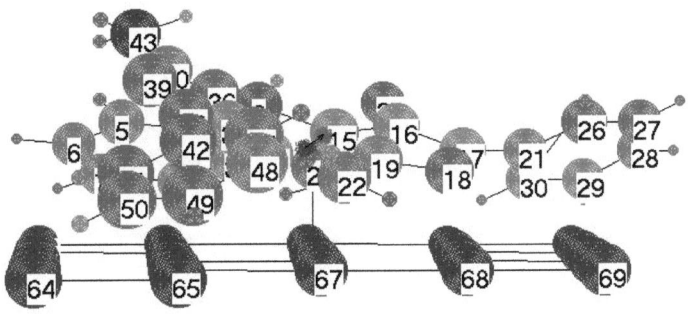

Fig. 7. **Differential** map showing **re-distribution** of electric charge in a molecule of oxyphenyl-diantipirylmethane **as a result of its adsorption** on Al_{20} model cluster (according to MNDO quantum chemical results). The colour scale as in Fig. 6.

Fig. 7 gives certain idea in illustrating the quantum chemical results showing **redistribution** of electric charge over the atoms of the molecule of the above said OIn at its adsorption on cluster modeling a fragment of Al surface, obtained by means of MNDO quantum chemical computations. Color illustrations are available in ref. [65], and they have been obtained with the use of visualisation software developed by the author in Borland Delphi 7.0 development media; the details of such ViSyMoDiD software were reported earlier in [58]. One of the features distinguising the author's ViSyMoDiD software [58, 65, 70] from similar visualisation programs (GaussView, ChemView, ChemDraw e.a. [62, 63]) is ability to choose (switch) from a view showing absolute net charges on atoms in a proposed colour scale to a **differential view showing the changes in atomic net charges** due to adsorption or any other changes occuring to a chosen molecular fragment a user may deem of; another feature is the possibility of showing a vector of dipolar moment in colour (pink semi-arrow at positive end and blue semi-arrom at negative end). On special 3D screen projection proposed and implemented, automatic rotation around any of the 3 principal axes can be either started or stopped independently of two other axes [58, 65, 70].

A correlation analysis of the results of Hartree-Fock *ab initio* quantum chemical computations and experimentally measured efficiencies of substituted sulphanilamides as inhibitors of corrosion and hydrogen absorption, was used to find ways of understanding the most likely explanations for the mechanisms of their action at the interface between steel and corrosion medium.

The mass-loss corrosion rate (CR) of mild steel was measured after 30 days exposure in aqueous salt media with substituted sulphanilamides (SS) at 1.0...10 mMol/L, and 1 of the 4 deuteromicetes (D): *Aspergillus niger (D1), Penicillum chrysogenum (D2), P. charlessii (D3), and PhialofOIn fastigiata (D4)*. Hydrogen absorption (HA) by corroded steel measured with the use of anodic dissolution technique corresponds to the corrosion rates. Both CR and HA are accelerated by the deuteromicetes, and their stimulating action is decreasing in the above listed order (1 to 4). The experimentally measured inhibitive efficiencies (IE) on CR (listed in Table 10 below) are analysed together with the obtained quantum chemical results using pair correlation technique.

Table 10

Corrosion Inhibiting Efficiencies of the Series of N-containing Organic Compounds at Presence of Deutemomicetes

Deuteromicete:	*Aspergillus niger (D1)*			*Penicillum chrysogenum (D2)*		
SS, mMol/L	1	2	5	1	2	5
IE (SS1), %	68	77	94	64	81	95
IE (SS2), %	44	60	78	56	75	86
IE (SS3), %	84	89	94	55	75	84
IE (SS4), %	82	86	94	52	40	81

As seen from the Table 11 below, there is steady correlation between the efficiencies of SS against corrosion and certain quantum chemical characteristics (HOMO and LUMO energies of the free SS molecules, Mulliken net charges on the atoms that comprise the active sites of SS molecules).

Table 11

Correlation Coefficients between the Computed Parameters of SS Isolated Molecules and Anti-Corrosion Inhibitive Efficiencies

Pair correlations (%) between IE and:	*(D1)*, 1 mMol/L	*(D1)*, across all SS conc.	*(D2)*, across all SS conc.	*(D3)*, across all SS conc.	*(D4)*, 2 mMol/L	*(D4)*, across all SS conc.
HOMO	44	33	− 48	− 5	− 84	− 47
LUMO	47	33	− 36	− 64	− 64	− 40
Q (S)	− 76	− 55	39	68	69	31
Q (O)	− 49	− 36	39	72	76	44
Q (N)	− 70	− 51	40	69	71	34
Dipole moment	70	53	− 44	− 81	− 89	− 42

The present work continues a series of publications [13 – 24] introducing our approach in understanding mechanisms of inhibiting action of N-containing organic compounds on both corrosion and hydrogen absorption of steel. The quantum chemical data compared in Table 11 with the experimentally measured efficiencies of corresponding inhibitors listed in Table 10, were obtained using Gaussian software [25].

Results shown in Table 11 suggest that different action mechanisms are involved in inhibitive action for different deuteromicetes: dipole moment calculated for the SS molecules, shows in general stronger correlation with EI than other tested quantum chemical characteristics (energies of highest occupied molecular orbital HOMO that characterises electron donating properties of the molecules [26], energy of lowest unoccupied molecular orbital LUMO which characterises electron accepting molecular properties, and net Mulliken atomic charges on heteroatoms of the molecular system studied – S, O, N). Even if we take, again, just the dipole moment, then it is possible to see that the sign of the pair correlation coefficient is different for D1 compared to the rest of deuteromicetes (D2, D3, D4). Quantum chemical data were obtained not only for the isolated molecules, but also for the individual SS molecules each adsorbed on model metal cluster containing 6 Fe atoms thus representing metal (steel) surface.

2.2. Study of acetylides

Efficient inhibitors of hydrogen absorption by steel can be selected on the basis of the known structure-properties relations [23, 27]. To make such relations known, existing quantum chemistry methods can be used. In the present report such molecular characteristics as dipole moment and energies of boundary orbitals (HOMO and LUMO) obtained by means of PM/3 [60, 61] quantum chemical method are compared to (a) biocide action against sulfate reducing bacteria (SRB) of a series of acetylides, and (b) inhibitive efficiencies of a series of ureides against hydrogen absorption by steel. The analysis of the obtained data shows that it is merely dipole moment that well explains inhibitive action against hydrogen absorption by steel, however it is HOMO that is rather responsible for describing biocide action of organic species against SRB.

Table 12

Structures of acetylides studied by the authors of [29]

Title	Structural Formula
A1	
A2	
A3	
A4	
A5	

In paper [29], acetylides (bicycles A1, A2 and tetracycles A3–A5, see Table 12) have been studied experimentally as SRB biocides. Such biocydes are important in protection of steel equipment primarily at oil extraction sites against corrosion and hydrogen absorption (the latter otherwise leading to deterioration of mechanical properties of construction materials, first of all, of steels). The corrosive media in which the SRB culture was developing as referred to in publication [29] was formation water from one of actual oil wells on the territory of Azerbaijan.

PM/3 has been chosen as the most appropriate quantum chemical methods as it combines simplicity (hence reasonable computer time consumption) of semi-empirical methods with precision of forecasting energies (as well as charge distributions in molecules on the basis of which integral characteristics such as dipolar momentae can be suitably predicted). Commercially available software [25] was used for performing quantum chemical calculations. The summary of the most important PM/3 results for A1–A5 molecules is presented in Table 13.

Table 13

Integrated results from PM/3 quantum mechanical computations

Title	HOMO Energy, β	LUMO Energy, β	Dipole Moment, D
A1	−0.27	−0.06	1.82
A2	−0.27	−0.05	1.69
A3	−0.22	−0.05	5.66
A4	−0.28	−0.08	4.68
A5	−0.27	−0.07	5.25

Highest occupied molecular orbital HOMO and lowest unoccupied (vacant) molecular orbital LUMO together are so called boundary orbitals; they characterise the 1st ionization potential of the whole molecule under question and its electron affinity respectively [26, 31].

The data on efficiencies of studied species against SRB (hereinafter referred to as **ESRB**) from [29] are reproduced herebelow in Table 15.

In Table 14, the line corresponding to 'A3' species is missing because addition of that substance lead to increase of bacteria growth hence 'A3' was found in the experiment [29] as being inefficient against SRB (in contrary to four other A's screened by the authors of [29]). In other words, the presence of 'A3' stimulates SRB growth in the media studied. The results presented in Table 13 were compared to the experimental results obtained in [29] by means of pair correlation analysis. The correlation coefficients resulting from the aforementioned comparison are brought in Table 15.

Table 14

Efficiencies against sulphate reducing bacteria (SRB) of A1–A5 species from the experiment [29]

Titles / concentrations	0.1 g/L	0.2 g/L
A1	70.2	75.5
A2	56.5	54.3
A4	43.0	45.6
A5	71.0	69.8

Table 15

Pair correlation coefficients (%) between microscopic and macroscopic (ESRB) properties resulting from analyzing together the data from Tables 13 and 14

Microscopic properties (listed below)	Concentrations of species in experiment [29]	
	0.1 g/L	0.2 g/L
HOMO Energy	**86**	**76**
LUMO Energy	39	30
Dipole Moment	−16	−19

As seen from Table 15, the module of pair correlation coefficients is increasing with the concentration of the experimentally studied species only in case of dipolar moment; as for the energies of boundary orbitals (HOMO and LUMO), respective correlations between theoretical and experimental data diminish with concentration of the additions to the media chosen by the authors of [29] as the media in which the SRB culture was developing.

The relationships between microscopic properties of molecules and macroscopic properties of corresponding substances have been studied by using PM/3 quantum mechanics application within Gaussian software following the previous studies [24, 28]. Since the purpose of this study was to examine the relationships between microscopic properties of molecules and macroscopic properties of corresponding substances using quantum mechanics, it was possible by means of existing Gaussian software [25] to perform PM/3 quantum mechanical computations and find the following new facts about the above mentioned relationship: pair correlation between HOMO energy and **ESRB** is strong (0.76 … 0.86, the corresponding figures are bold highlighted in Table 15), while the one of LUMO and **ESRB** is much weaker (so that correlations are insignificant for that case).

This means that the 1st ionization potential (characterized by HOMO energy) of A1–A5 molecules (as a whole) strongly (and positively) correlates with the anti-SRB activity of corresponding 'A' species (within the selected A1–A5 row) unlike electron affinity of 'A' molecules (LUMO energy). As for dipole moment, its relationship with **ESRB** macroscopic property turns to be extremely weak (meaning that for the selected species dipole moment nearly does not influence the **ESRB** property). The negative sign of those correlation coefficients means that increase of dipolar moment in our row of A1–A5 molecules tends to even decrease anti-SRB efficiency. PM/3 method was found to be rather successful in pointing out the prevalence of donor mechanism of biocide action (against SRB) for the studied species above acceptor mechanism [70, 78].

2.3. Study of ureides

The same quantum chemical method, PM/3 was used in the present study as a means to find the relationship between quantum chemical descriptors as above for A1–A5 (based on the molecular structures of the studied species) and another macroscopic property of corresponding species, i.e. inhibiting efficiency of organic substances U1–U9 (added in experiments performed elsewhere [32] to the corrosive media) against hydrogen absorption. In Table 16, the molecular structures of the studied ureides U1–U9 are listed sequentially together with corresponding molecular weights.

The respective species have been assigned titles U1…U9. PM/3 method did converge at default parameters only for U1, U2, U6 and U8 (but did not for the rest of the 'U' series). The following Table 17 lists the values of quantum chemical descriptors of the individual molecules constituting those compounds. Chosen properties are, as above for 'A' series, dipole moments, as well as HOMO and LUMO energies.

In table 18 are quoted some experimental data obtained elsewhere in a certain experiment [21] in which current density varied from 1.5 up to 9 Amp/dm^2. Experiment [21] involved measurements of efficiency of each of the organic additions U1–U9 against hydrogen absorption by skin layers of steel (such absorption is very unfavorable for mechanical properties of materials because it results in deterioration of plasticity hence eventually in cracking of structures made of such materials [4]). In fact, the efficiencies listed Table 18 represent retaining the plasticity measured on steel wire specimen.

Table 16

**Ureides studied by means of PM/3 method as inhibitors
against hydrogen absorption by steel**

Title	Structural Formula
U1	CH —— CO —— NH$\diagdown$ $\quad$ CH $\diagdown$ $\parallel$ $\qquad\qquad$ C ==== S CH —— CO —— NH$\diagup$
U2	CH$_2$ —— CO —— NH$\diagdown$ $\qquad$ $\vert$ $\qquad\qquad$ C ==== S CH$_2$ —— CO —— NH$\diagup$

U3	O—CO—CH₃ ... S—C=N—CO—CH₂ / NH—CO—CH₂ ... *HCl ... O—CO—CH₃
U4	Br, Br substituted quinone S—C=N—CO—CH₂ / NH—CO—CH₂ *HCl
U5	Br, Br substituted quinone S—C=N—CO—CH₂ / NH—CO—CH₂ *HCl
U6	OH, Br, Br, Br substituted S—C=N—CO—CH₂ / NH—CO—CH₂ *HCl OH
U7	O—CO—CH₃, Br, Br, O—CO—CH₃ substituted S—C=N—CO—CH₂ / NH—CO—CH₂ *HCl
U8	OH, Br, Br substituted S—C=N—CO—CH—Br / NH—CO—CH—Br *HCl OH

The data on efficiencies of a series of 7 ureides against corrosion of mild steel are listed in Table 19. These data [27] are based on experimental results obtained gravimetrically in [25]. The concentrations of the organic inhibitors (OIn) added to Postgate B standard media were 1, 2, and 10 mMol $\cdot$ dm^{-3}; the corrosion rate at absence of OIn was $k_0 = 3.5$ g/(m^2 $\cdot$ day) used as a reference when calculating the inhibiting efficiencies (IE) for each of Inh $Z = (k_0-k)/k_0$ (1).

Table 17

The values of quantum chemical descriptors for 'U' molecules obtained by means of "PM/3" computations

TITLES	QCC from PM/3 quantum chemical computation		
	Dipole moment (D)	HOMO (β)	LUMO (β)
U1	3.99	-0.35	-0.07
U2	4.74	-0.34	-0.06
U6	1.37	-0.34	-0.05
U8	9.08	-0.34	-0.07

Table 18

Efficiencies of the chemical species (EIHA)
corresponding to 'U' molecular structures against hydrogen absorption by steel

Organic additions to the media in experiment	Macroscopic efficiencies of 'U' species obtained for cathode current densities D_c (Amp/dm^2)			
	1.5	3	6	9
Reference data (no additions)	58	56	55	50
U1	66	65	62	57
U2	67	66	63	60
U6	69	68	64	61
U8	78	76	72	65

Table 19

Efficiencies of ureides (ECI) [27] based on corrosion rates [40]
for various concentrations of the studied corrosion inhibitors

Inhibitor number	Inhibitor concentration		
	1 mMol $\cdot$ dm^{-3}	2 mMol $\cdot$ dm^{-3}	10 mMol $\cdot$ dm^{-3}
	Inhibiting efficiencies, %		
U1	20	29	34
U2	26	29	40
U3	43	54	71
U4	37	51	69
U5	40	49	60
U6	26	31	43
U7	34	43	57

To compare the experimental results from [40] with our quantum chemical data, the following descriptors have been chosen: energies of the boundary molecular orbitals (HOMO and LUMO) and dipole moment. *Ab initio* computations (in a basic set 6-31G*) as well as calculations obtained by means of three semi-empirical methods (MNDO, PM/3, AM/1) were used with the aid of commercially available software [25]. The results of pair correlation analysis of those IE experimental results together with our quantum chemical data presented in Table 20 show that dipole moment and HOMO significantly correlate with IE while LUMO data do not show any strong correlation. HOMO characterizes 1st ionization potential of a molecule analysed by means of MO LCAO methods while LUMO stands for affinity to an electron. Table 20 is based on PM/3 results because the data obtained by 3 other methods tested, has lead to generally much less significant **r**'s.

Table 20

Correlation coefficients (r, %) between the values of selected QCD (from Table 17) and ECI (from Table 19) of the studied ureides

Quantum Chemical Descriptors:	Concentrations of OIn, $mMol \cdot dm^{-3}$		
	1	2	10
Dipole Moment	**−67**	**−79**	**−83**
HOMO	**−84**	**−88**	**−84**
LUMO	−29	−28	−12

Within the framework of the previously developed approach, it means that the inhibitive properties of the selected ureides under chosen conditions are based on electron donating rather than accepting properties of their molecules. Also, the significance of dipole moment of the molecules of OIn for explaining protective mechanism of organic OIn against corrosion, has been re-confirmed under given conditions [28].

PM/3 method is preferable compared to other methods tested (i.e., MNDO, AM/1), and even non-empiric *ab initio* methods because it is found to be more helpful in explaining the mechanism of inhibitive action of organic species.

Results shown in Table 19 represent high values of pair correlation coefficients (77…82 %) between dipole moment and ExD for all experimental current densities except for the highest one (9.0 Amps/dm^2). Those values are shown in Table 19 in **bold**.

In case of boundary orbitals, neither HOMO and LUMO demonstrate significant correlation with ExD except for D_c=9.0 Amps/dm^2 (and only in case of HOMO, not LUMO). This means that boundary orbitals responsible for donor-acceptor interaction do not play any important role in explaining the studied macroscopic property of the chosen ureides (U1–U9). However, the consistency of the very sign of those correlation coefficients for HOMO (r>49 %) and LUMO (r< –3 %), as seen from Table 24, tells us that the values of HOMO energy taken for different organic species chosen are basically in direct dependence with experimentally established [32] efficiency of respective species, and respectively LUMO energy is in inverse relationship to the same **ExD**.

Table 21

Pair correlations between EIHA experimental results reproduced above in Table 18 and the theoretical results summarized above in Table 17

D_c values from experiment (Amp/dm^2)	Correlation coefficients (%) between **experimental data** and theoretical results from PM/3 quantum chemical computations		
	Dipole moment	HOMO	LUMO
1.5	**78**	49	−32
3.0	**77**	50	−30
6.0	**82**	47	−36
9.0	64	**76**	−3

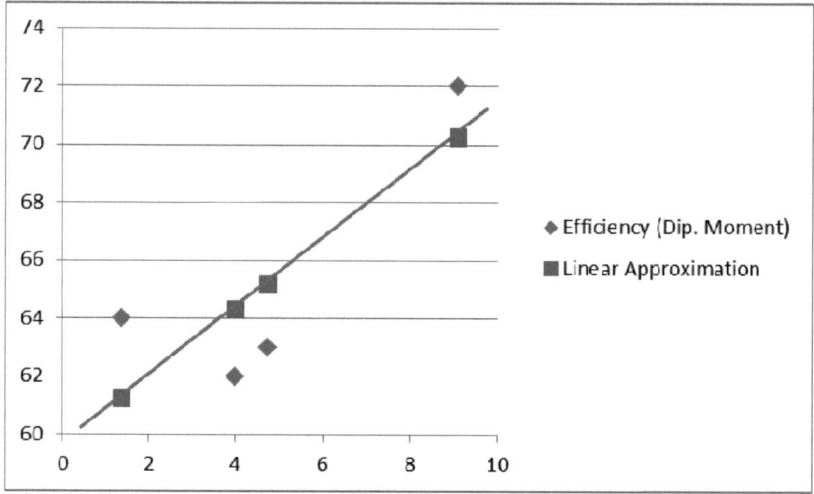

Fig. 8. Efficiencies of ureides (ExD, ordinates) against hydrogen absorption by steel (at 3 Amps · dm^{-2}), linearly approximated as a function of dipolar moment (abscissa).

3. Quantum chemical study of morpholine derivatives as inhibitors of sulfide corrosion at the presence of SRB

In paper [33] eleven morpholine derivatives R_1 $-C_6H_4$ $-SO_2$ $-N*$ (R_2) $-CH_2$ $-$ CHOH $-NC_4H_8O$ where R_1 and R_2 are listed in table 22, have been experimentally studied as inhibitors of corrosion of mild steel in water salt media imitating sea water inoculated by SRB. The efficiencies of inhibitive anti-corrosion action (ECI) of the OIn have been determined in [33] (the experimental data are reproduced in table 22 of the present work). The authors of [33] have underlined that heterocyclycity of morpholine and its derivatives favors adsorption [4,34–37] of the studied OIn on the surface of metal being protected.

Table 22

Efficiencies of OIn of corrosion of mild steel according to the data published in [33]

OIn	R_1	R_2	ECI, %
M1	p–F		23
M2	p–Cl	$-CH_3$	-1
M3	p–Br		5
M4	$m, m, p-$ $(CH_3)_3$		42
M5	p–Cl		40
M6	p–Br	$-C_4H_9$	53
M7	p–I		61
M8	p–F		48
M9	p–Cl	$-C_6H_5$	56
M10	p–I		60
M11	$m, m, p-$ $(CH_3)_3$		72

To refine adsorption (and eventually protection) mechanisms of OIn on theoretical level, quantum chemical computations of corresponding isolated molecules have been carried out using PM/3 method. PM/3 method has been chosen as a compromise between complicated but precise *ab initio* method consuming sometimes enormous time for multi-atomic systems and acceptable trustworthiness/quality of the obtained results which is characteristic for semi-empirical methods.

Quantum chemical calculations of the isolated molecules named above in Table 22 as M1–M11 have been performed using Gaussian software [25]; the basic results are brought together in Table 23. Usually protective action of OIn (particularly, their adsorption characteristics) are related to the values of dipole moment, energies of boundary orbitals (HOMO, LUMO) [31, 26], and Mulliken net charges on atoms (primarily, on heteroatoms) of the molecular systems considered [23, 24, 41–43].

In table 24, ECI from [33] have been compared to the results of quantum chemical calculations (see Table 23). From Table 25 it appears that under the studied conditions of corrosion, the most significant are pair correlation coefficients between ECI of the studied OIn and such QCD of their molecules as dipole moment ans Mulliken net charges on N, O, S heteroatoms.

Table 23

The values of the selected quantum chemical descriptors
(according to PM/3 computations)

OIn	Quantum chemical descriptors (QCD)					
	HOMO energy (β)	LUMO energy (β)	Dipole moment (D)	Charge on N* ($|e|$)	Charge on O ($|e|$)	Charge on S ($|e|$)
M1	−0.353	−0.044	1.252	−0.170	−0.264	2.391
M2	−0.352	−0.039	1.254	−0.209	−0.305	2.426
M3	−0.352	−0.044	1.245	−0.171	−0.264	2.373
M4	−0.324	−0.039	1.796	−0.171	−0.265	2.340
M5	−0.350	−0.025	1.359	−0.173	−0.260	2.378
M6	−0.349	−0.044	1.297	−0.170	−0.255	2.381
M7	−0.349	−0.042	1.289	−0.162	−0.261	2.371
M8	−0.286	−0.044	1.629	−0.089	−0.265	2.282
M9	−0.328	−0.043	2.410	−0.126	−0.259	2.360
M10	−0.332	−0.038	2.296	−0.125	−0.257	2.348
M11	−0.333	−0.021	2.043	−0.132	−0.271	2.348

The results presented in Table 24 may be explained by the fact that the charges on N*, S, and O heteroatoms are characterizing adsorption properties of the studied OIn. The positive values of the pair correlation coefficients between ECI and the net Mulliken charges on O and N* mean that the best of the eleven OIn studied are those with less negative charges on O and N* atoms while the negative value of correlation between ECI and the net charge on S atom (varying from one OIn molecule to another) means that among the OIn studied, the ones with less positive charge on S atom are more efficient.

Assuming that in accordance to Kupmans theorem, HOMO energy corresponds to the first ionization potential of a molecule while LUMO – to its electron affinity [38,39], within the studied row of OIn there are neither pure donators nor pure acceptors. It favours the concept that for the studied OIn, adsorption of the mixed type takes place.

Table 24

Results of pair correlation analysis
of experimental [33] data together with quantum chemical data

QCD	Pair correlation coefficients (%) between ECI and QCD
HOMO energy	38
LUMO energy	33
Dipole moment	**60**
Charge on N*	**64**
Charge on O	**59**
Charge on S	−54

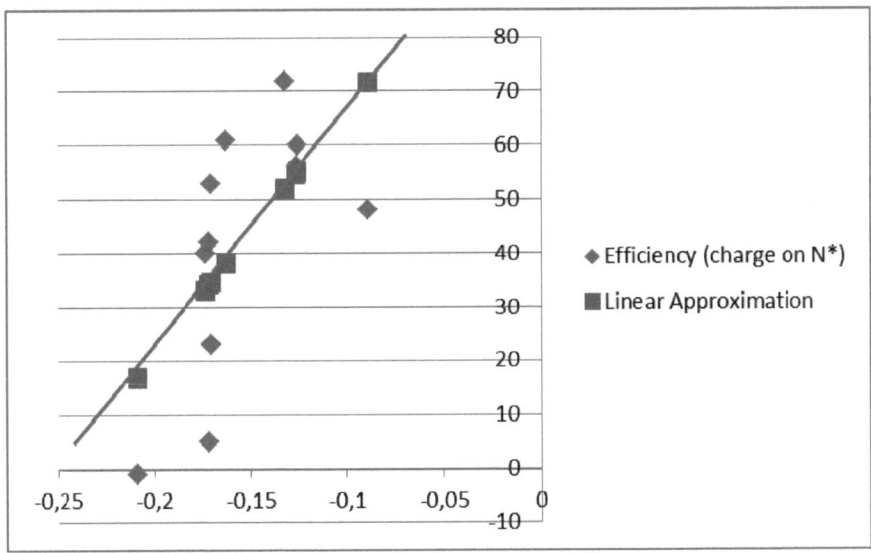

Fig. 9. Efficiencies of morpholine derivatives (ordinates) as linear approximation of net Mulliken charges on nitrogen atom designated above as 'N*'.

The most significant pair correlation coefficients were found to be between experimental data [33] on efficiencies of the inhibitors against corrosion in water-salt media, on one hand, and computed dipolar moments of their isolated molecules, as well as net Mulliken charges on N, O, S atoms, on another.

The present conclusions are in favour of the hypothesis that the mechanism of adsorption of the inhibitors (the derivatives of morpholine) is based on adsorption of inhibitors on the surface of the metal being protected against corrosion. Neither donor nor acceptor action prevails for the studied series of morpholine derivatives.

4. Study of industrial herbicydes as inhibitors of corrosion of metals

Industrially synthesized herbicydes of the general formula:

$$R_1 - C9 - N5 - R_2,$$

(where R_1 =H$_2$ CH$_3$, H$_2$ C$_2$H$_5$, H (CH$_3$)$_2$; R_2 =N NH CCl N NH C$_2$H$_5$) proved to be also corrosion inhibitors. The experimentally determined ECI data on corrosion of Cr-Ni steel in *Postgate B* media are listed in Table 25. PM/3 quantum chemical method was used for analysis of OIn molecules by means of [25] software; the results of computations are given in Table 26.

Table 25

Experimentally measured ECI of a series of industrial herbicydes
(experimental conditions: corrosion of Cr18Ni10 steel in *Postgate B* media)

OIn titles	ECI Z (%) at OIn concentrations:				
	1 mMol/L	2 mMol/L	5 mMol/L	10 mMol/L	15 mMol/L
Г1 (carbation)	40	45	48	51	55
Г2 (pitezin)	23	34	47	66	82
Г3 (prometrin)	23	34	47	66	82
Г4 (symazine)	19	38	43	55	76
Г5 (зеазин)	21	30	45	57	78

Computed data (Table 26) were compared to the experimental results (Table 25) in Table 27 by means of pair correlation analysis. From Table 27 it appears that electric net charge on C9 atom practically does not correlate with ECI (except for the case where concentration in corrosion media is 5 mMol/L). At small concentrations, high positive values of pair correlation coefficients are obtained between ECI and dipole moment of OIn molecules, HOMO and LUMO energies, where at high concentrations studied – high negative values. As for net atomic charge on the Nitrogen atom, on the opposite, at small concentrations of OIn the corresponding pair correlations possess high negative values, whereas for high concentrations **r** values are high positive.

38

Table 26

Results of PM/3 quantum chemical computations
of the molecules of industrial herbicydes

Titles	HOMO energy, a.u.	LUMO energy, a.u.	Dipole moment p, Debye	El. charge on N_5 atom, $|e|$	El. charge on C_9 atom, $|e|$
Г1	– 0.137	0.146	10.5	– 0.018	– 0.297
Г2	– 0.363	– 0.046	1.46	0.877	– 0.385
Г3	– 0.335	0.012	0.56	0.907	– 0.290
Г4	– 0.365	– 0.047	1.42	0.885	– 0.410
Г5	– 0.364	– 0.047	1.33	0.884	– 0.412

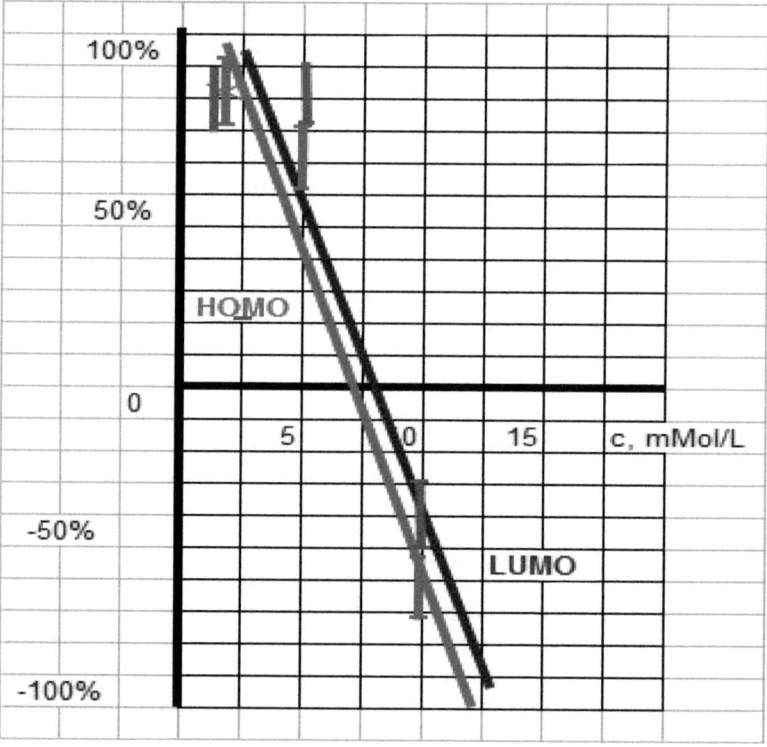

Fig. 10. Concentration dependencies of pair correlation coefficient between ECI and HOMO, ECI and LUMO for a series of herbicydes showing the features of corrosion inhibitors.

Table 27

Pair correlation coefficients **r** (%) between the ECI values determined form experiment (from Table 25) and QCD (from Table 26)

QCD	r (%) for OIn concentrations:				
	1 mMol/L	2 mMol/L	5 mMol/L	10 mMol/L	15 mMol/L
HOMO energy	**99.7**	85.1	75.3	− 55.7	− **94.6**
LUMO energy	**98.3**	81.9	83.9	− 39.8	− 87.2
Net el. charge on N5 atom	− **98.2**	− 85.9	− 66.2	67.4	**98.3**
Net el. charge on C9 atom	65.7	47.3	**89.2**	26.6	− 36.8
Dipole moment	**97.0**	86.1	61.9	− 71.6	− **99.2**

5. Study of antipyrine, pyrimidyne, and pyrroline derivatives

To predict inhibitive efficiency of an organic compound, thus avoiding redundant costly organic synthesis making it a purposeful one, one needs correlation analysis of protective efficiencies together with quantum chemical results. *Ab initio* computations were performed with the use of Gaussian software. Experimental efficiencies (measured for 8 diantipirylmethane derivatives, OIn) are reported in Table 28 below. Notations "Cor1" through "Hab10" are designated for purpose of Table 30.

Table 28

Experimentally measured efficiencies of studied OIn

Efficiencies at concentrations (mMol/L)	Against corrosion, %			Against hydrogen absorption, %		
	1	5	10	1	5	10
Notation:	*Cor1*	*Cor5*	*Cor10*	*Hab1*	*Hab5*	*Hab10*
OI1	45	66	76	26	43	50
OI2	53	70	78	54	56	59
OI3	67	75	79	28	54	57
OI4	61	69	78	23	51	53
OI5	40	68	70	13	47	48

The resulting pair correlation coefficients are listed in the table 29 where $\Sigma Q(C)_i$ denotes sum of Mulliken charges on benzene ring. Non-significant correlations are truncated in Table 29. Results give evidence of the mixed type of inhibitive action [1–3].

Anaerobic corrosion is one of the most dangerous types of microbiological corrosion, the sulphate reducing bacteria (SRB) being the main exciters of such corrosion. Certain organic substances perform double action, both as corrosion inhibitors and bactericides (possibly even being first of all bactericides and thus indirectly inhibiting corrosion).

Table 29

Internal Pair Correlations (%) between various Quantum Chemical Results

LUMO	94						
ΔE	− 98	− 99					
Charge (N)	98	89	− 94				
Charge (O)	− 55	− 53	54	− 68			
ΣQ(C)$_i$	− 96	− 85	91	− 96	56		
Charge (C*)	17	94	− 12	17	12	4	
Dipole moment **p**, D	− 97	− 54	91	− 96	47	90	− 38
QCD	HOMO	LUMO	ΔE	Q (N)	Q (O)	ΣQ(C)$_i$	Q (C*)

Table 30

Correlation Coefficients between Quantum Chemical Data Obtained for the Isolated Molecules of OIn using Software [25], and Experimentally Measured Efficiencies of Corresponding OIn against Corrosion

QCD and notations (see Table 12)	*Cor1*	*Cor5*	*Cor10*	*Hab1*	*Hab5*	*Hab10*
HOMO	76	91	-	-	-	-
LUMO	82	88	71	-	-	64
ΔE	81	90	62	-	-	57
Q (N)	63	89	-	-	-	-
Q (O)	-	-	-	-	-	-
ΣQ(C)$_i$	− 60	− 75	-	-	-	-
Q(C*)	-	-	-	-	79	-
p, D	− 75	− 94	-	-	-	-

Herebelow the results of our study of diantipirylalkanes as inhibitors of corrosion and hydrogen absorption by steel CrNi 1810 are reported together with the simultaneous evaluation of their bactericide action against SRB. The experimental measurements were proceeded using Postgate B media at 37^0C, corrosion rate measured gravimetrically. The following parameters were controlled during exposition of the specimen: pH and E_h of the media, concentration of biogenic H_2S in it, SRB bacterial titer and electrode potential of the steel specimens. Spectrophotometer SF-46 was used to measure the distribution coefficient (WHDC) for the studied OIn molecules between the two non-interacting liquids (water and n-hexane that models bacterial membrane). After 48 hours of exposition, the specimens made of stainless steel together with OIn were put into the media containing the developing SRB [5,7,9,10].

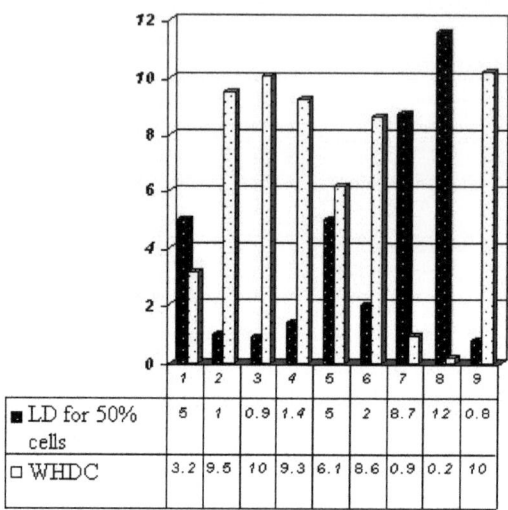

Fig. 11. Relation between LD_{50} and WHDC.

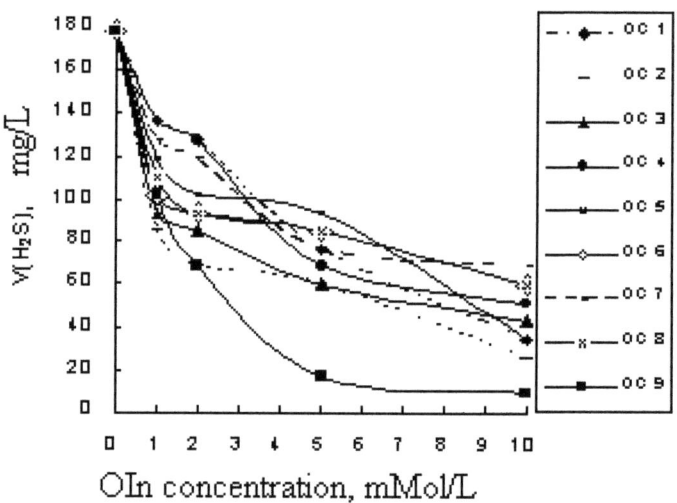

Fig. 12. Dependence of H_2S content on OIn concentration in the media.

Table 28 is based on internal correlations [18] between the quantum chemical results of the theoretical study, and suggests that either $\Delta E=$LUMO–HOMO or, for example, LUMO results be excluded due to the fact of their strong inter-correlation

(r $=-$ 99%). Computed dipole moment **p** of isolated OIn molecules strongly correlates with HOMO energy ($r = -97\%$) and charge on N atom of OIn ($r = -96\%$).

The addition of the studied OIn into the corrosive media results in the suppression of SRB activity. The reverse dependence between 50% Lethal Dose (LD_{50}) and the above mentioned (aforementioned) WHDC has been estimated (see Fig.11). Inhibiting efficiency of the studied OIn against microbiological corrosion reaches 70...80 %, and by hydrogenation of steel – 50...70 % at OIn concentration 10 mMol L^{-1}. All studied OIn efficiently suppress bacterial sulphate reduction, minimal H_2S content in the closed anaerobic system reached at OIn concentration 10 mMol L^{-1} , see Fig. 12. Some of the important experimentally obtained dependencies giving additional understanding of the proposed models of inhibitive action [22, 23] are shown as Fig. 13... Fig. 18.

Our results give evidence of the mixed type of inhibitive action. In the case of corrosion, mechanism of electron donation by OIn represented by HOMO [26] line in Table 29 prevails upon electron accepting action of OIn represented by LUMO [26] line of the same Table.

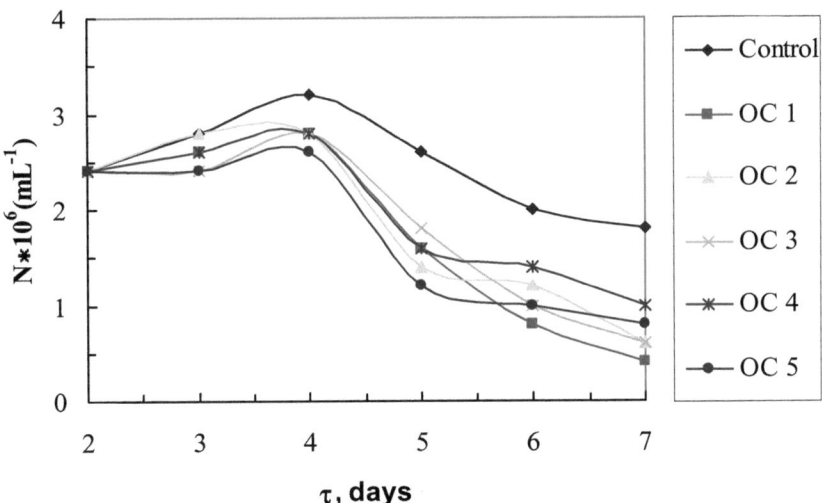

Fig. 13. Number of SRB cells as a function of exposure time in aqueous salt corrosion medium in the presence of organic compounds at concentration 5 mM·L^{-1}.

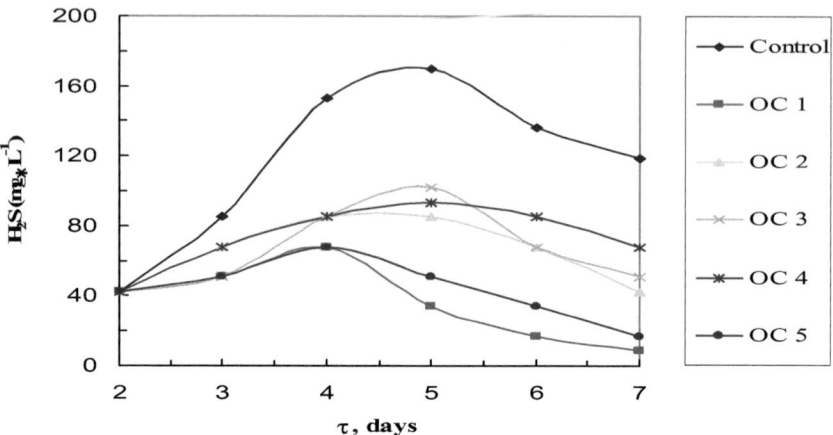

Fig. 14. Changes in the values of hydrogen sulfide concentration with exposure time for the system "Stainless steel-Postgate B medium with **SRB**" in the presence of organic compounds at concentration 5 mM·L^{-1}.

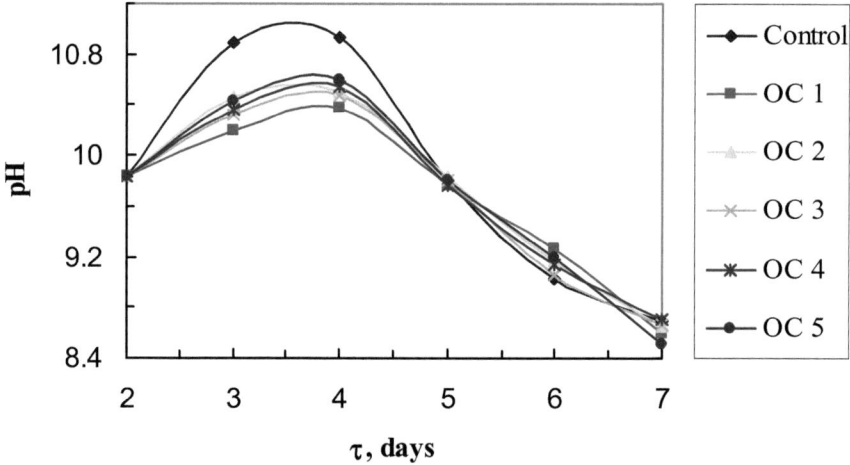

Fig. 15. Changes in the pH values with exposure time for the system "Stainless steel-Postgate B medium with **SRB**" in the presence of organic compounds at concentration 5 mM·L^{-1}.

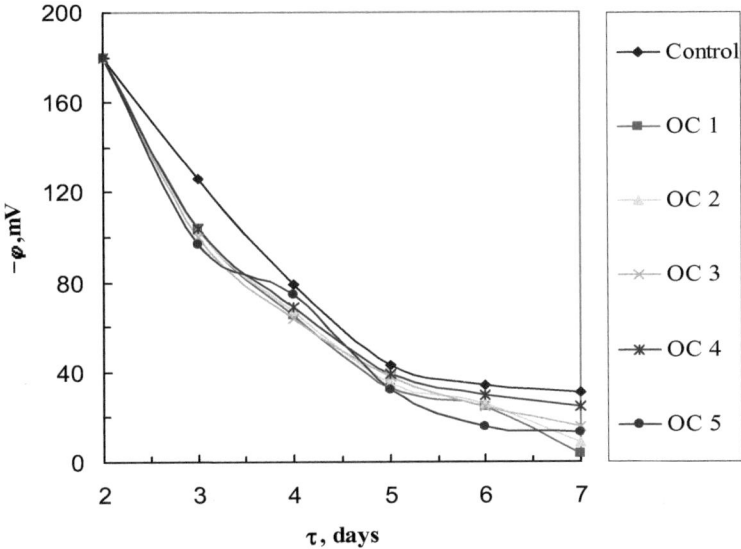

Fig. 16. Changes in the values of electrode potential of steel specimens with exposure time for the system "Stainless steel-Postgate B medium with **SRB**" in the presence of organic compounds at concentration 5 mM·L^{-1}.

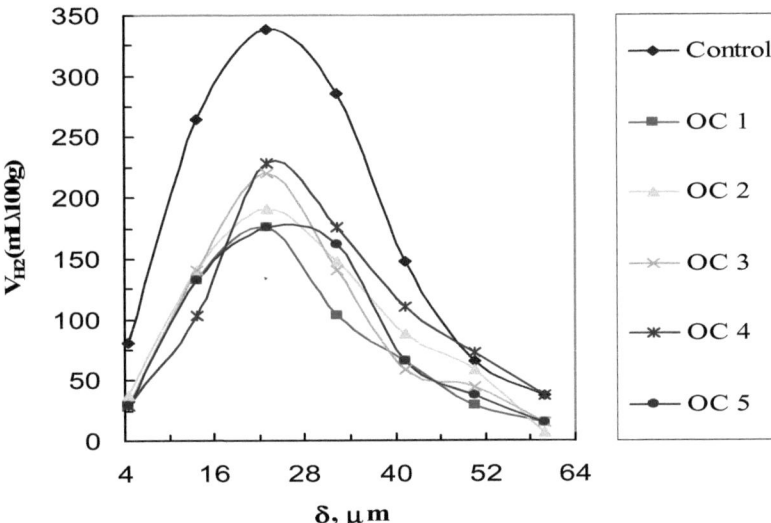

Fig. 17. Hydrogen distribution across the depth of stainless steel specimens (electrolyte contains organic compounds at concentration 5 mM·L^{-1}).

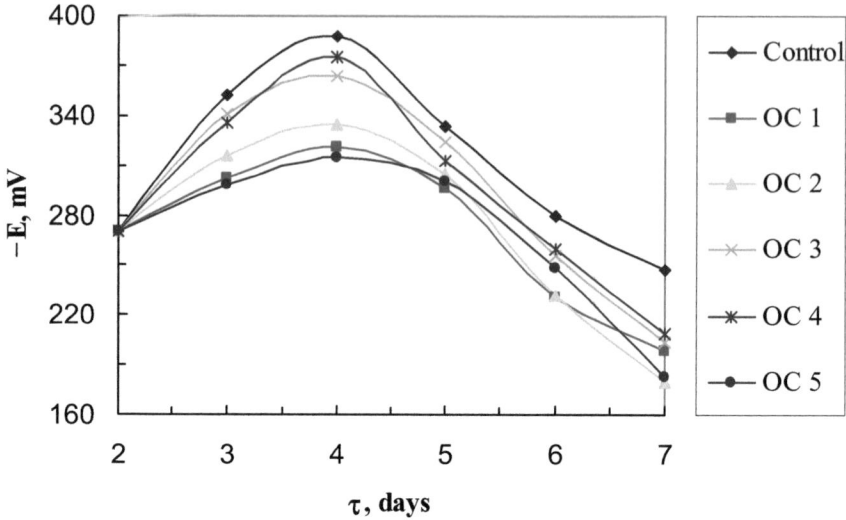

Fig. 18. Changes in the values of redox potential exposure time for the system "Stainless steel-Postgate B medium with **SRB**" in the presence of organic compounds at the concentration of 5 mM·L^{-1}.

It is shown that in case of hydrogen absorption by steel, only electron accepting properties (characterised by LUMO energy) of OIn studied as well as their molecule charge distribution in the area of certain carbon atom marked above as C*: r (Hab5)=79%, are important from the pOInt of OIn inhibiting efficiency.

On the contrary, in case of corrosion, such characteristics shown for Q(C*), is obviously not important in understanding the inhibitive action, while the other three – namely, net Mulliken charge on nitrogen as certain active point of the OIn Q(N), sum of Mulliken charges on carbon forming benzene ring fragment as the molecule active site $\Sigma Q(C)_i$, and dipole moment p of isolated molecule – seem essential ($r = 89\%$, $r = -75\%$, and $r = -94\%$ respectively at OIn concentration c=5 mMol/L) in the framework of the concepts of molecular orbitals and donor-acceptor interaction between adsorbent and adsorbate.

In [64] the author has compared ECI and EIHA data with QCDs for the following series of OIn: [N(1)CCH$_3$–CH–C(6)OH–N(2)–C(5)] N(3)H–N(4)=R, где R: =CHCH$_3$, =CHCH=CHCH$_3$, =C(C$_2$H$_5$)$_2$, =C(–CH$_3$)(–C$_3$H$_7$), =C(–CH$_3$)(–C

$(CH_3)_3$), $=C(-CH_3)(-CH_2-C$ $(CH_3)_3$), $=C(-CH_3)(-C_6H_{13})$, $=CH-C_6H_4-m-OH$ synthesised by Laima Salickaite [41] in Vilnius. The numbers in brackets stand for reference in Table 31 below. Steel samples were subjected to hydrogen absorption by cathode polarization at current density of $D_k= 0.02$ Amp/cm^2 for 48 minutes. Water electrolyte referred in Table 29 as «EL1» contained 0,1 n H_2SO_4, and its only difference to «EL2» was the addition of 5 mg/L of H_2SeO_3 (as promoter of absorption of cathodically evolved hydrogen [4, 28, 52]).

Table 31

Efficiencies of Cr18Ni10 steel (%) inhibitors titled as C1–C8 in the two electrolytes: EL1 – 0.1 n H_2SO_4 solution, EL2 = EL1+ 5 g/L of H_2SeO_3

Inhibitor:	C1	C2	C3	C4	C5	C6	C7	C8
Electrolyte								
EL1	62	100	39	77	69	69	62	23
EL2	62	92	46	77	77	77	54	60

In Table 32, the pair correlation coefficients between ECI (from Table 31) with QCDs computed for isolated molecules of the corresponding species using MNDO method and software [25].

As seen from Table 32, pair correlation between LUMO energy with ECISRB (OIn efficiency against corrosion at presence of SRB) is characterised by rather high positive values of the coefficient **r** (90 %) while correlation between HCMO and ECI (protecting efficiency in sterile media) – with some smaller but also positive **r** value (37 %).

Assuming also the big negative value of pair correlation coefficient **r** between HOMO energy and ECISRB (– 93 %) and the negative **r** value of much smaller magnitude obtained between HOMO and ECI (– 43 %), it is possible to conclude that on the basis of performed quantum chemical computations of the molecules of pyrimidyne derivatives [41] together with pair correlation analysis of QCC experimental data on E and (Table 32), at adsorption of the molecules of the chosen class of IC/IH on the steel surface donor-acceptor interaction takes place, which is in this case both "Mechanism A" donating an electron (or pair) from metal to OIn molecules, and, at the same time, donating an electron (or pair) by OIn moleculs in favor of the metal ("Mechanism B").

Table 32

Pair correlation coefficients **r** (%) between QCD and EIHA (by Cr18Ni10 steel) or anti-corrosion efficiency of the pyrimidyne derivatives (C1–C8) at corrosion of the same steel in sterile (ECI) and SRB-inoculated media (ECISRB)

QCD	EIHA	ECISRB	ECI
E_{HOMO}	− 32	**− 93**	− 43
E_{LUMO}	8	90	37
ΔE	25	92	41
Q(N1)	20	**93**	44
Q(N2)	− 40	**− 98**	− 57
Q(N3)	− 43	**93**	44
Q(N4)	− 33	− 92	− 40
Q(C5)	− 16	**93**	44
Q(C6)	50	53	97
Sum of net Mulliken charges on –CH group	− 28	− 83	− 23

For antipyrine derivatives the most significant correlation coefficients (varying from 68% up to 89%) were obtained between LUMO energies and ECI; correlation coefficients for EIHA were of higher values.

On the contrary, for pyrroline derivatives the best correlation was obtained for between ECI and HOMO energies (from 88% to 92%), see Table 32 above. As for EIHA and HOMO, correlation was less strong (67…74 %).

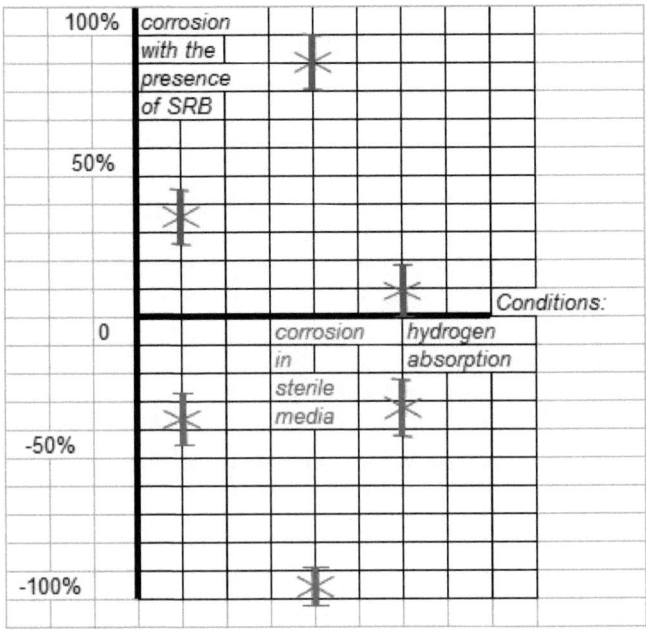

Fig. 19. Pair correlation coefficients r(c) between HOMO (—) and LUMO (—) values, on one hand, and efficiencies of studied inhibitors against: (a) corrosion at the absence of SRB, (b) corrosion in presence of SRB, and (c) – hydrogen absorption by steel, on another.

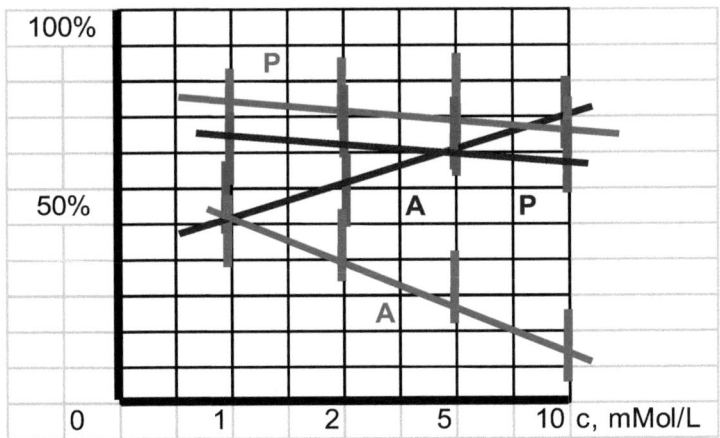

Fig. 20. Concentration dependencies of pair correlation coefficients between ECISRB (at presence of SRB) and the computed HOMO (—) and LUMO (—) energies for OIn – antipyrine (A) and pyrroline (P) derivatives.

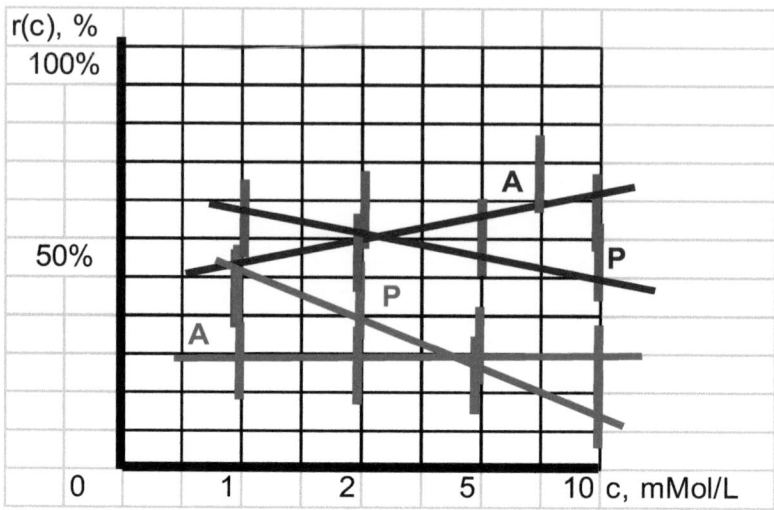

Fig. 21. Concentration dependencies of pair correlation coefficients between EIHA and the computed values of HOMO (——) and LUMO (——) energies for inhibitors – antipyrine (A) and pyrroline (P) derivatives.

The results show that at adsorption of OIn molecules of antipyrine derivatives on surface of mild steel, donor-acceptor interaction takes place presumably **donating an electron (or pair) by metal in favor of OIn molecule**. Such donation appears to be more significant in case of corrosion inhibition (Fig. 20, 'A' elements) rather than hydrogen absorption (Fig. 21, 'A' elements).

Performed quantum chemical computations show that at adsorption of OIn molecules of pyrroline derivatives on surface of mild steel, donor-acceptor interaction takes place presumably **donating an electron (or pair) by OIn molecule to metal**. Such donation is more significant in case of corrosion inhibition rather (Fig. 20, 'P' elements) than hydrogen absorption (see Fig. 21, 'P' elements).

Thus, it is possible to conclude that **adsorption both of antipyrine and pyrroline derivatives features chemisorption rather than physical adsorption, and also, depending on chemical structure of OIn molecules, for such chemisorption on metal (which is steel in both of the cases) molecule there is an alternative: electron shift may prevail either in the directon from metal to OIn molecule (presumably, electrons from LUMO) or opposite, from OIn molecule towards metal (presumably, electrons from HOMO).** Then it is clear that **preferential displacement of electron cloud from OIn admolecules towards the atoms of** metal leads to formation of double electric layer on metal / solution interface, its positive charge side oriented towards the solution, thus electrostatically repelling H_3O^+ ions.

6.1. CONCLUSIONS

Permeability of adsorption layer of OIn molecules for discharging H_3O^+ ions or H_2O molecules (if one understands by this arising opportunity of emerging hydrogen adatoms on the surface of metal), depends on its coverage θ by OIn admolecules and bond strength between them and metal surface. The coverage defines lifetime of OIn molecule on the surface. Such permeability depends not only on the nature of metal and solution composition and temperature but also on composition and structure of organic molecules which has been proven by abundant experimental studies held during the last third of 20th century [4, 28], including the ones performed by the author [7, 73, 74].

Coverage θ is the function of atomic packing density on metal faces which determines electron work function hence adsorption properties if one does not focus on the influence of special stressed deformed state of solid metals (associated with the presence of implicit layer of crystallites deformed to various degrees) and other physico-chemical properties of metal on atomic packing density function. It is natural that coverage θ depends on bond strength between OIn molecules and surface atoms of metal.

From past experimental data associated either (at least) partly or as a whole to inhibiting corrosion of metals and hydrogen absorption by steel, it is possible to conclude that:

(1) for OIn to be efficient, it should possess active sites such as polar OH, CO, SO_2NH_2 e.a. groups, N, S and O heteroatoms;

(2) non-saturated chemical bonds such as double (– C=C, C=N, see Fig. 22) and/or aromatic would be an advantage [28, 42] for OIn molecules to be more efficient. Only in very few publications [18, 20, 42], attempts to interpret experimental results by means of attracting the concept of boundary orbitals (HOMO, LUMO), absolute electronegativity η, so called «absolute rigidity» [31] of molecules $\Delta E/2$ have been made so far;

(3) presence of sufficiently big dipole moment [67, 69] of a molecule provides with efficient adsorption of such OIn molecules hence sound inhibitive action.

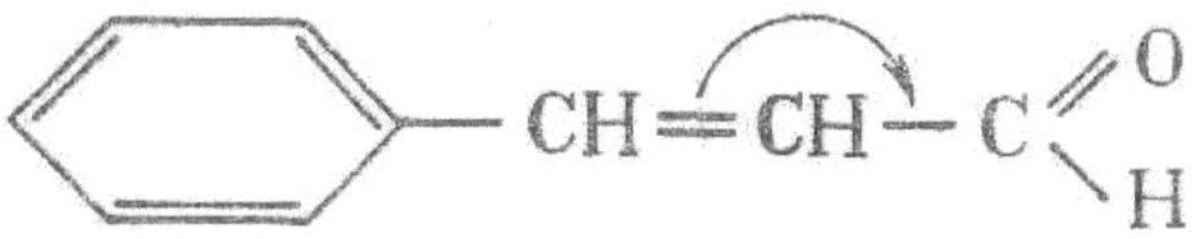

Fig. 22. Induction effect in a molecule of cinnaroic aldehyde [28] that shows best inhibitive efficiency against hydrogen absorption by steel of the studied aldehydes.

The general conclusions presented below support the hypotheses about the possible mechanisms of inhinbitive action of organic species, particularly the one related to significant improvement of inhibitive efficiency at increasing negative charge on O atom in alkylaromatic OIn molecules with double bonds near carbonyl C due to a shift of electron density from the double bond to Oxygen atom of carbonyl group [28].

Table 33

The overall list of the seria of studied inhibitors of corrosion of metals, of hydrogen absorption by steel, and inhibitors-biocydes

Group title of series	Respective chemical class
F, Г	Substituted phenols and herbicydes
E	Diethylamineacetic acid derivatives (substituted amines and amides)
D	Diantipyrine derivatives
P	Pyrroline derivatives
S	Pyrimidyne derivatives
A	Acetilydes
M	Morpholine derivatives
U	Ureides

The overall results of the undertaken studies within the framework of approaches to explaining mechanisms of inhibitive action of the species listed in Table 33, are presented in Table 34.

Individual group titles defined by Table 33 are shown in Table 34 on condition that for such series of species, there are significant pair correlations **r** between respective QC D and ECI; «+» signs denote positive and «–» – negative[1] significant **r**. An absence of certain group titles in a table cell means there are no significant correlations **r** available for the given QCD and OIn series under the given

---

[1] «*» in a designation stands for the cases where the sign of corresponding correlation coefficient **r** changes once (one time) at monotonous increase of concentration.

conditions. In Table 34, the designations presented in parentheses denote the adsorbent metals studied, while the ones in square brackets characterize the concerned active site of OIn molecules; text in figure brackets either refers to other tables or contains comments.

As seen from Table 34, in case of corrosion inhibition, <u>positive</u> significant pair correlations **r** between **ECI and HOMO** energies are exhibited <u>by «D», «P», and «Г» groups</u> (refer to Table 33), For the latter («Г»), the sign changes once (i.e., single time) at monotonous increase of OIn concentration). <u>Negative</u> significant pair correlation **r** is between HOMO and ECI <u>for «S» group</u>.

As for correlations between **ECI and LUMO,** <u>positive</u> **r** is obtained for «D» and «S», while the negative **r** value is for «Г». Again, in the case of «Г» the sign of **r** changes once (i.e., one time) at monotonous increase of OIn concentration. **It means that an increase of such QCD values as «HOMO energy» are associated with increase of ECI for «D», «P», and «Г» listed above while in case of «S» group it would generally lead to decrease of ECI (see Table 32).**

Table 34

Overview analysis of the roles of the chosen QCDs in explaining ECI and EIHA mechanisms along the seria of the OIn studied (referred to as the group titles from Table 33)

QCD S/N	The values of computed QCDs	**ECI**	**EIHA**
1	HOMO energy	+A, +D, +P, −S, −U, −Г*	+P, +U
2	LUMO energy	+S, +D, −Г*	+D
3	Dipole moment **p** of isolated molecule of OIn	+D, +M, +P, −U, −Г*, −F {subset}	+U
4	Net electric charges Q_i on active molecular sites of OIn	+E[N]+E[C_6H_4] −E[CH_3], +D[N], −P* {Table 33}, +S[C] +S[N1] −S[N2], −D[C_6H_n], +M[N], +M[O]	−P, +D[C]
5	**«Differential polarities»** of individual bonds Δq, refer to (6), (7)		+E

6	Absolute electronegativity of OIn molecule η, see (4)	−E	
7	Absolute rigidity of OIn molecule $\Delta E/2$, see (3)	+D, +E, +S	+D, −E
8	Most probable angle α between dipole moment vector of OIn admolecule modified due to adsorption and the normal to metal surface plane	−E, −F	
9	Total electric charge ΣQ on admolecule obtained due to its adsorption	+D(Cd)	
10	Change of the magnitude of dipole moment Δp of adsorbed OIn molecule due to its adsorption on model cluster	+F(Al), −D(Cd)	

Positive significant **r** was obtained between ECI and dipole moment **p** (refer to Table 32) for OIn of «M», «D» and «P» groups, while negative – for «F» and «Г» (in case of «Г»), the sign of **r** changes once at monotonous increase of OIn concentration. Between ECI and net electric charge(s) Q_i on atoms or fragments, **r** is positive in cases of «E» group (on N atom and on aromatic ring), «D» group (on N), and «S» group (for N1 and carbon atoms permanently present in the whole series) but negative for «E» group (on methyl fragment), «D» group (for aromatic ring), «S» group (for N2 permanently present in the whole series, but not on N1, see above), «M» group (for N, O atoms) and «P» group (however with a single sign change as discussed above). Taking into consideration that larger electronic density corresponds to more negative charges on atoms, it is negative **r** that points to direct relationship between ECI and electron density on atom or atomic fragment.

Apart from that, negative significant **r** has been obtained between ECI and η for OIn of «E» group, and positive – between ECI and ΔE for «E», «D», and «S» groups; negative significant **r** was obtained between ECI and α angle for OIn of «E» and «F» groups, positive – between ECI and ΣQ for «D» group (in case of its OIn model adsorption on Cd surface). Also, positive significant **r** has been obtained between ECI and the **absolute change of dipole moment due to adsorption Δp modelled by means of quantum chemistry** for OIn of «F» group (in case of adsorption on Al), negative – for «D» group (on Cd).

Thus, for corrosion OIn of various structure, the analysed QCDs suggest possible existing differences in interaction mechanisms of OIn molecules with metal surface as the adsorbent. In case of inhibition of **hydrogen absorption** by steel, positive significant pair correlation coefficients **r** between <u>EIHA and HOMO</u> are demonstrated <u>by OIn of «U» and «P» groups</u> (see Tables 33, 34), and also between <u>EIHA and LUMO</u> energies – for <u>«D» group</u>. Negative significant **r** was obtained between EIHA and Q_i for OIn of «P» group, while positive for – «D» group (namely, in case of carbon atom permanently present in that whole group), and also positive significant **r** values for «E» group – between EIHA and Δq which is the change of net atomic charge (on an atom permanently present in the whole group) due to adsorption. Moreover, positive significant **r** has been obtained between EIHA and ΔE for OIn of «D» group, while negative – for «E» group.

To compare the mechanisms of protection by OIn in cases of corrosion of metals and hydrogen absorption by steel, let us compare the ECI and EIHA columns in Table 31: synbate behaviour of pair correlation coefficients **r** takes place in cases of such QCDs as «HOMO energy» (but only for «P» group of OIn), «LUMO energy» (but only for «D» group), «Absolute rigidity of OIn molecule» (for «D» group, but not for «E» or «S»), and also for the net charges Q_i on active molecular sites, if one compares ECI to EIHA only within the «P» group.

It is interesting that for the QCD titled «Absolute rigidity of OIn molecule», the signs of **r** for OIn of «E» group (Table 34) are opposite for the cases of inhibition of corrosion and of hydrogen absorption, i.e.:

(a) for this «E» group of OIn, the influence of that QCD on inhibition of corrosion is of opposite character to inhibition of hydrogen absorption.

From the analysis of data presented in Table 31, it follows that two more factors make evidence of the different inhibition mechanisms in the above said 2 cases.

Factor (b): for «E» group, significant negative correlation between ECI and absolute electronegativity η takes place while as for inhibition of hydrogen absorption, EIHA correlation with another QCD – «Differential polarity of a individual chemical bond in OIn molecule» such correlation is positive (and by the way, from Table 9 it can be seen that this is about C–N bond).

Factor (c): the relationship of individual Mulliken net charges Q_i on active molecular sites of OIn with ECI is different from the one with EIHA: i.e., for «D» group of OIn when inhibiting hydrogen absorption, the concept of electron density on permanent carbon atom is of importance to be responsible for protective mechanism (positive significant **r** brings the evidence of direct relationship between EIHA and positive charge on that atom, but at the same time it is necessary to keep in mind that positive charge corresponds to less electron density), whereas at corrosion for «D» group the significant **r** values are the ones between EIHA and the net electric charge distribution characteristics for other active sites of OIn molecules, namely, net Mulliken charge on permanently present N atom and the sum of net charges on the carbons in aromatic ring. The analysis of the remaining **r** values presented in line 4 of Table 31 for other series of OIn, has been made above. The above mentioned differences of the results of choosing the QCDs according to their significance of corresponding correlation coefficients, in case of hydrogen absorption from the case of corrosion, are probably related to much more important role of surface defects (exits of dislocations, grain boundaries, and other defects of crystalline lattice [75]) to be considered in case of hydrogen absorption compared to conditions indispensable for consideration in case of corrosion inhibition.

To analyse the differences in selection of QCDs on the basis of the importance of such QCD for explaining the biocyde action of OIn against SRB, from the case without such biocyde action, we compare the last column in Table 31 with the last but one – let us take pyrimidyne derivatives as the example («S» group). The only significant **r** in case of inhibiting corrosion at presence of SRB is obtained for Q(C6), i.e. ECI due to the OIn of «S» group is as more as the net charge on atom designated in the general «S» group structure (on page 38) as «C(6)» is more positive.

It is interesting that in the cases of corrosion without SRB and hydrogen absorption, the QCD titled «Net charge on C(6) atom» is, on the contrary, not significant (refer to table 31), while the other 9 descriptors listed in Table 34 (including the ones which relate to the net charges on any of the 4 Nitrogen atoms permanent for the whole series) are significant form the point of corrosion inhibition in media without SRB but are significant for neither inhibiting corrosion with SRB nor inhibiting hydrogen absorption by steel.

6.2. SUMMARY

The target quantum chemical descriptors (this term has been suggested in [12, 44, 59, 70]) for purpose of optimisation of molecular structure of OIn are the QCDs most significantly correlating with experimental data inhibition of corrosion or/and absorption of cathodically evolved hydrogen, and may be obtained by means of semi-empirical quantum chemical computations such as PM/3, MNDO, restricted Hartree-Fock *ab initio* [25, 31, 39, 60, 61, 76, 77]. Such QCDs selected for single molecules of OIn as dipole moment, energies of boundary orbitals (HOMO, LUMO), net Mulliken charges on individual atoms of active molecular sites of OIn. At the same time, none of the listed QCDs is appropriate as an individual one to be able to provide with *a priori* selection of the most efficient OIn from less efficient [70].

Also, it is shown above that the QCDs obtained at quantum chemical modelling of adsorption of OIn molecule on emulated surface of substrate are also useful for purpose of explaining mechanisms of inhibiting action in the framework of the adsorption approach. Such QCDs are: the change of magnitudes of dipole moments of OIn molecules due to their adsorption on surface of metal being protected, changes of net atomic charges on individual atoms of active sites as well as on such individual fragments as a whole (i.e. summed up by their atoms), the value of the total electric charge appearing on adsorbed molecule due to interaction with a substrate [70, 72, 73, 74].

The magnitude of dipole moment computed for individual OIn molecules, significantly correlates with ECI, EIHA, and biocyde efficiencies (where applicable) for all 9 groups of the studied species.

Performed quantum chemical modelling shows that both the magnitude and orientation of the vector of dipole moment **p** in intramolecular coordinates are changed at adsorption of, for example, substituted phenols on either Al, Al_2O_3 or $Al(OH)_3$ model surfaces so that even the scalar Δp value itself exhibits significant correlation with ECI of such substituted phenols [67, 70].

Optimal cluster size responsible for correctly modelling the properties important for adsorption of OIn on surface of protected metal at the state of the art level of development of computer facilities is found to be from 8 to 25 substrate atoms (depending on how large in dimensions and complicated are the admolecules of OIn proposed for such modelling) [67, 70].

The optimal orientations of the molecules of susbstituted phenols at their adsorption on modelled Al_2O_3 and $Al(OH)_3$ correspond to the range of the angles between the vector of dipole moment of adsorbed OIn molecule and the normal to the plane of substrate surface from 46 ° to 129 °. As for pure Al surface also used for such quantum chemical modelling, the orientations of molecules of substituted phenols at their adsorption correspond to some wider range of the optimal angles between the vector of dipole moment of adsorbed OIn molecule and the normal to the plane of substrate surface, namely from 24 ° up to 135 °[67, 70].

7. REFERENCES

1. I.L. Rosenfeld. Corrosion inhibitors. – 1977. – Moscow, "Khimia". 351 p. (*in Russian*).

2. L.I. Antropov, E.M. Makushin and V.F. Panasenko. Metal corrosion inhibitors. – 1981, Kiyev, "Tekhnika". – 184 p. (*in Russian*).

3. S.M. Reshetnikov. Inhibitors of acid corrosion of metals. –1986, Leningrad, "Khimia". – 144 p. (*in Russian*).

4. S.M. Beloglazov. Electrochemical hydrogen and metals: absorption, diffusion, and embrittlement prevention in corrosion and electroplating. – 2011, New York, Nova Science Publishers, Inc. – 260 p.

5. S.M. Beloglazov, G.S. Beloglazov, and A.A. Myamina. Proc. 46th Ann. Meeting Int. Soc. of Electrochem., Xiamen, China, 1995.

6. S.M. Beloglazov, G.S. Beloglazov. 3rd European Federation of Corrosion 'Workshop on Microbial Corrosion', Estoril Portugal, 1994. – Abstr. – P. 64.

7. S.M. Beloglazov, N.N. Gorilenko, G.S. Beloglazov. Corrosion in natural and industrial environments: Problems and industrial solutions, Grado, 1995. P. 211–218.

8. Z.R. Agayeva. Development of inhibitors of corrosion of metals on the basis of N- and S- containing organic species and products of utilization of wastes and by-products of petrochemical industry // Thesis. Dr. Sci., Chem. – Baku, 2013 (*in Russian*).

9. S.M. Beloglazov, I.A. Ermakova, E.M. Kondrasheva and L.V. Malashenko. Proc. VI Korosyon Semposyumu Bildiriler Kitabu, Ankara, Turkey, 1998. – P. 337–342.

10. S.M. Beloglazov. Advances in Materials Science, 2007, V. 7. – P. 202–209.

11. B.B. Damaskin. Electrochemistry, 2011, V. 47. – P. 1058–1065 (*in Russian*).

12. V.I. Vigdorovitch, L.E. Tsygankova. Inhibiting hydrosulhuric and carbon-dioxide corrosion of metals. – 2012. Moscow, Kartek. – 244 p. (*in Russian*).

13. S.M. Beloglazov, G.S. Beloglazov. Conference Papers Eurocorrosion-94, Bournemouth, UK, 1994. – V. 3. – P. 94 – 100.

14. G.S. Beloglazov, S.M. Beloglazov. Quantum chemical study of nitrogen and sulphur containing substances as inhibitors of the corrosion and hydrogenation of steel // Development in Marine Corrosion. – 1998, Cambridge, The Royal Soc. Chem. P. 143 – 154.

15. G.S. Beloglazov, S.M. Beloglazov, M.V. Gryaznova: Proc. EMCR'2006. – Nice, France, 2006. – P. 37.

16. G.S. Beloglazov, M.V. Gryaznova, S.M. Beloglazov. Proc. EUROCORR 2004. – Nice, France, 2004, CD ROM, 6.

17. G.S. Beloglazov, A.A. Myamina, S.M. Beloglazov. EUROCORR 2004. – Abstr. Nice, France, 2004, P. 39.

18. G.S. Beloglazov, M.V. Gryaznova, S.M. Beloglazov. 55th Annual Meeting of ISE. – Thessalonica, Greece, 2004.– Vol. 2. – P. 942.

19. S.M. Beloglazov, G.S. Beloglazov, M.N. Laykova: Proc. 10 Europ. Symp. 'Corros. and Scale Inhib. SEIC', Ferrara, Italy, 2005, Sez. 5, Suppl. 12, P. 505–516.

20. G.S. Beloglazov, S.M. Beloglazov. Quantum chemical study of adsorption of organic on Al and Cd surfaces // Achievements and prospects in development of new medicinal drugs: Proc. Russ. Sci.-Pract. Conf... 70 Anniv. PSPhAcad., Perm, Russian Federation, 2007, P. 163–167 (*in Russian*).

21. G.S. Beloglazov, S.M. Beloglazov and O.D. Gladysh. Progress in the Understanding and Prevention of Corrosion. – Proc. 10th European Congress, Barcelona, 1993. – P. 900–905.

22. D.M. Malyarevsky, S.M. Beloglazov. Advances in Materials Science. – 2007. – V. 7. – P. 216–223.

23. S.A. Teryusheva, G.S. Beloglazov, and S.M. Beloglazov. Experimental and quantum chemical study of quinone derivatives as inhibitors of corrosion and hydrogen absorption by steel // Advances in Materials Science. 2011. V. 11. No. 3 (40). – P. 42–49.

24. S.A. Teryusheva, G.S. Beloglazov, S.M. Beloglazov. Experimental and quantum chemical study of quinone derivatives as inhibitors of corrosion and hydrogen absorption by steel // Solid State Phenomena (ISBN 978-3-03785-320-7). – 2012.

V. 183. – P. 249–255.

25. Gaussian 09, Revision A.02, M.J. Frisch, G.W. Trucks, H.B. Schlegel, G.E. Scuseria, M.A. Robb, J.R. Cheeseman, G.Scalmani, V.Barone, B.Mennucci, G.A. Petersson, H. Nakatsuji, M. Caricato, X. Li, H.P. Hratchian, A.F. Izmaylov, J. BlOIno, G.Zheng, J.L.Sonnenberg, M. Hada, M. Ehara, K. Toyota, R. Fukuda, J. Hasegawa, M. Ishida, T. Nakajima, Y. Honda, O. Kitao, H. Nakai, T. Vreven, J.A. Montgomery, Jr., J.E. Peralta, F. Ogliaro, M. Bearpark, J.J. Heyd, E. Brothers, K.N. Kudin, V.N. Staroverov, R. Kobayashi, J. Normand, K.

Raghavachari, A. Rendell, J.C. Burant, S.S. Iyengar, J. Tomasi, M. Cossi, N. Rega, J.M. Millam, M. Klene, J.E. Knox, J.B. Cross, V. Bakken, C. Adamo, J. Jaramillo, R. Gomperts, R. E. Stratmann, O. Yazyev, A.J. Austin, R. Cammi, C. Pomelli, J.W. Ochterski, R. L. Martin, K. Morokuma, V.G. Zakrzewski, G.A. Voth, P. Salvador, J.J. Dannenberg, S. Dapprich, A.D. Daniels, O. Farkas, J.B. Foresman, J.V. Ortiz, J. Cioslowski, D.J. Fox, Gaussian, Inc., Wallingford CT, 2009.

26. M.A. Landau. Molecular Mechanisms of Action of Physiologically Active Compounds, 2nd edn, 323 p.; 1979, Moscow (*in Russian*).

27. A.A. Sikachina, G.S. Beloglazov, S.M. Beloglazov. Study of Ureides Derivatives as Inhibitors of Corrosion of Mild Steel in Salt Aqueous Media at Presence of SRB // EUROCORR2013. Estoril, Portugal. – Proc.

28. S.M. Beloglazov. Hydrogen absorption by steel at electrochemical processes. – Leningrad, State university press. – 1975. – 412 p. (*in Russian*).

29. A.E. Smorodin, N.M. Agayev e.a. Suppression of sulphate reducing bacteria by means of cyclic acetylene derivatives // Protection of Metals. 1983. No. 3. P. 471– 473 (*in Russian*).

30. C. Hansch, T. Fujita. p-σ-π Analysis. A method for the correlation of the biological activity and chemical structure // JACS, 1964. – V. 86. – P. 1616.

31. V.F. Traven. Electronic Structure and Properties of Organic Molecules. – M.: Chemistry. – 1989. – 384 p. (*in Russian*).

32. S.M. Beloglazov, K.V. Egorova, N.V. Kolesnikova. Control of microbial corrosion and hydrogen absorption by steel plated in the bath with organic inhibitors // EUROCORR-2002: Internat. Symp. – Extended Abstr. on CD. – Riva del Garda. – 2001. – Abstr. No. 100.

33. N.N. Gorilenko, S.M. Beloglazov. Influence of Morpholine derivatives on biogenic sulphide corrosion of steel // Corrosion and protection of metals. – 1985. – P. 481 – 483 (*in Russian*).

34. S.M. Beloglazov, N.N. Gorilenko, G.S. Beloglazov. Corrosion in natural and industrial environments: Problems and industrial solutions, Grado, 1995. P. 211.

35. S.M. Beloglazov, G.S. Beloglazov. Abstr. 3rd European Federation of Corrosion Workshop on Microbial Corrosion, Estoril Portugal, 1994. Abstr. 64.

36. S.M. Beloglazov, I.A. Ermakova, E.M. Kondrasheva, L.V. Malashenko. Proc. VI Korosyon Semposyumu Bildiriler Kitabu, Ankara, Turkey, 1998. P. 337.

37. S.M. Beloglazov. Advances in Materials Science, 2007. Vol. 7, P. 202.

38. G.S. Beloglazov, S.M. Beloglazov, A.Yu. Tretyakov. Study of organic nitrogen-containing inhibitors of corrosion and hydrogen absorption of metals using PMR method // Spectroscopy of coordination species: Proc. 5th All-union Conf. – Krasnodar, October 24-30, 1988. – P. 71. (*in Russian*).

39. H.H. Dunken, Ch. Opitz. Zu einem einfachen LCAO-MO-Modell fuer die Chemisorption einfacher Molekuele an einigen Metallen: Elektronenverteilung und Valenzschwingung // Z. Chem. – 1966. – B. 6. – No. 10. – S. 390 – 391.

40. N.V. Kolesnikova. Influence of ureides derivatives on electrodeposition of alloy Ni-Mn and its corrosion at presence sulfate reducing bacteria and absorption of hydrogen. Thesis. Kaliningrad. 2004 (*in Russian*).

41. L. Salickaite. Synthesis and study of hydrazinepyrimidines derivatives // Thesis. Chem. – Vilnius, Vilnius university. 1985 (*in Russian*).

42. S.A. Teryusheva., S.M. Beloglazov, G.S. Beloglazov. 1,4-hydroquinones as inhibitors of corrosion and hydrogen absorption by structural steel in media with sulfate reducing bacteria // Practical Anti-Corrosion Protection // 2008. No. 4 (50). P. 60 – 65 (*in Russian*).

42. S.A. Teryusheva, G.S. Beloglazov, S.M. Beloglazov. Quantum chemical modeling of the mechanism of action of inhibitors of corrosion and hydrogen absorption by steel based on the derivatives of 1,4-quinone at the presence of sulphate reducing bacteria // Proc. Samara State univ. Series: Natural Sci. 2011. Issue 2 (83). P. 184 – 189 (*in Russian*).

43. S.A.Teryusheva, G.S. Beloglazov, S.M. Beloglazov. 1,4-Benzoquinone derivatives as inhibitors of corrosion and hydrogen absorption of steel at presence of SRB // Proc. Samara State univ. Series: Natural Sci. 2011. Issue 5. P. 138 – 145 (*in Russian*).

44. A.J. Stuper, W.E. Bruegger, P.C. Jurs. Computer assisted studies of chemical structure and biological function // John Wiley & Sons. – 1979.

45. P.F. Cox, R.L. Every, O.L. Riggs. Study of aromatic amine inhibitors by nuclear magnetic resonance // Corrosion. – 1964. – V. 20. – No. 9. – P. 299 – 302.

46. Ya.M. Potak. Brittle fractures of steel and steel parts. Moscow, 1955. – 452 p. (*in Russian*).

47. V. Dress. Vacuum deposition of metal coats // Iron Age. – 1957. – V. 180. – No. 25. P. 142 – 145.

48. Postgate J.R. The Sulphate-Reducing Bacteria. – 2nd Edition. – Cambridge: Cambridge University Press. – 1984. – 208 p.

49. Postgate J.R., Campbell L.L. Classification of Desulfovibrio species, the non-sporulating sulphate-reducing bacteria // Bacteriol. Rev. – 1966. – V. 31. – P. 732 – 738.

50. C.P. Slikter. Principles of Magnetic Resonance, 2nd ed., Springer Verlag, Berlin, 1978.

51. V.K. Voronov, R.Z. Sagdeyev. Basics of magnetic resinance. – Irkutsk, 1995. – 352 p. (*in Russian*).

52. M.N. Polukarov. Proc. of Perm State university. – 1935. – V. 1. – P. 12 – 16. (*in Russian*).

53. I.N. Putilova, S.A. Balezin, V.P. Barannik. Inhibitors of corrosion of metals. – Moscow, 1958. – 184 p. (*in Russian*).

54. B.M. Larkin, I.L. Rosenfeld. Correlation between donor ability of aliphatic amines and their efficiency as corrosion inhibitors according to the data of quantum chemical calculations // Protection of metals. – 1976. – V. 12. – P. 259 – 263 (*in Russian*).

55. B.M. Larkin, I.L. Rosenfeld. Quantum chemical study of the mechanism of interaction of nitrogen oxides and N-containing ions with the surface of iron // Protection of metals. – 1978. – V. 14. – P. 643 – 651 (*in Russian*).

56. B.M. Larkin, I.L. Rosenfeld. Study of mechanism of action of inorganic corrosion inhibitors using molecular orbilats method // Protection of metals. 1977. – V. 13. – P. 170 – 175 (*in Russian*).

57. B.M. Larkin, I.L. Rosenfeld. Study of the properties of chromate ione at adsorbed state on iron fragments // Protection of metals. – 1977. – V. 13. – P. 450 – 454 (*in Russian*).

58. G.S. Beloglazov. Computer software for 3D-visualization of polyatomic systems // Pharmacy & Health: Proc. International Conference. – Perm (Russian Federation), 2005. – P. 92 (*in Russian*).

59. A.S. Kabankin, L.I. Gabrielyan. Study of the relationship between hepatoprotective detoxifcating activity of species and their molecular structure using quantum chemical descriptors // Chemical Pharmaceutical Journal. – 2008. – V. 42. – No. 5. – P. 30 – 32 (*in Russian*).

60. T. Clark. A Handbook of Computational Chemistry. Wiley-Interscience, 1985.

61. J.B. Foresman, A. Frisch. Exploring chemistry with electronic structure methods: a guide to using *Gaussian*. Gaussian, Inc. – 1993. – Pittsburgh. – 270 p.

62. GaussView // Retrieved April 27, 2013 from the World Wide Web: http://www.Gaussian.com

63. ChemView // Retrieved April 27, 2013 from the World Wide Web: http://www.cambridgesoft.com

64. G.S. Beloglazov, S.M. Beloglazov. Quantum chemical study of nitrogen and sulphur containing substances as inhibitors of corrosion and hydrogenation of steel. Proc. of the 8th Europ. Symp. Corr. Inhibitors (8 SEIC), Ann. Univ. Ferrara, N.S./ Suppl. No. 10. – 1995. – P. 1251 – 1258.

65. **G.S. Beloglazov. ViSyMoDiD – software for visualizing polyatomic molecular systems with features of showing absolute and differential electron distribution for the sub-systems together with dipole moment vector and 3 independent automatic rotations.** Retrieved April 27, 2013 from the World Wide Web:

// http://www.visymodid.0fees.net

66. G.S. Beloglazov, M.I. Vakhrin, V.P. Polyudova. Comparison of the results of computating electronic structure of some inhibitors of acid corrosion using Hueckel and Hoffman molecular orbital methods with high resolution PMR data // Application of corrosion inhibitors in national economy: Proc. Sci-tech. Seminar. Chelyabinsk, Russian Federation, 04-07 June 1983. – P. 19 – 20.

67. G.S. Beloglazov, S.M. Beloglazov. Inhibiting action of substituted phenols against corrosion of Al in media with bacterial sulphate reduction // 55th ISE Annual Meeting of ISE. Thessaloniki, Greece. 2004 / Abstracts. – V. 2. – P. 857.

68. G.S. Beloglazov, S.M. Beloglazov. Experimental and quantum chemical study of adsorption and protective action of inhibitors of corrosion and hydrogen absorption by metals // Innovations in science and education – 2010: Proc. International Sci. Conf. – Kaliningrad. – 2010. – P. 297 – 299.

69. I.G. Medvedev. Correlation effects at adsorption of hydrogen on transition metals // Thesis (Author's Abstract). – Moscow, 1998. – 42 p. (*in Russian*).

70. G.S. Beloglazov. Experimental and quantum chemical approaches to explaining adsorption and protective action of inhibitors of corrosion of metals and hydrogen absorption by metals. – Lambert Academic Publishing. – ISBN 978-3-659-36440-2. – 2013. – 169 p. (*in Russian*).

71. M.P.H. Brongers e.a. Corrosion costs and preventive strategies in the USA // Retrieved March 18, 2010 from the World Wide Web: http://www.corrosioncost.com/pdf/main.pdf.

72. G.S. Beloglazov. Modelling of adsorption of molecules of organic substances acting as inhibitors of corrosion and hydrogen absorption by metals using quantum chemistry // Mathematical Physics and its Applications: 2nd International Conf. Samara. – 2010. – P. 350 – 351.

73. G.S. Beloglazov, S.M. Beloglazov, M. Mngereza. Quantum chemical computation of some hydrazides and hydrazones as inhibitors of corrosion and hydrogen absorption by steel // Environmental Degradation of Engineering Materials (EDEMET-2011): International Conf. – Gdansk, 2011.

74. G.S. Beloglazov, J. Elisadiki. Quantum chemical computation of organic inhibitors of corrosion on Al and Cd surface. // Environmental Degradation of Engineering Materials (EDEMET-2011): International Conf. – Gdansk, 2011.

75. N.V. Kardash, V.V. Batrakov // Protection of metals. – 2000. – V. 36. – No. 1. – P. 64 – 66 (*in Russian*).

76. L.-M. Andre, J. Ladik. Quantum mechanical studies of polymers: present stutus and perspectives. – 1980. – 256 p.

77. P. Saalfrank, J. Ladik, R.F. Wood, M.A. Abdel-Raouf, C.-M. Liegener. Ab initio cluster and band structure calculations on systems modeling La2CuO4. Effects of charge transfer between the different planes, Madelung potentials, doping and correlations // Physica C-superconductivity and Its Applications - PHYSICA C. – 1992. – V. 196. – P. 340 – 356.

78. G.S. Beloglazov, S.M. Beloglazov, J. Makangara. Quantum chemical approach to selection of efficient inhibitors of hydrogen absorption by steel and of biocydes against SRB. // Internat. Conf. Wear Processes 2012. – Proc. – Poland, Szczecyn. 2012.

8. TABLE of CONTENTS

Printed by Books on Demand GmbH, Norderstedt / Germany